The Consulting Engineer

C. Maxwell Stanley
Stanley Engineering Company

THE CONSULTING ENGINEER

Presenting the professional and management problems involved in the practice of consulting engineering.

John Wiley & Sons, Inc., New York · London

Library of Congress Catalog Card Number: 61-11182

Printed in the United States of America

Preface

The consulting engineer performs an important role in the American economy and industry. He renders a useful and valuable service both to governmental and to corporate organizations. Unfortunately, his services and responsibilities are poorly understood.

The public at large cannot distinguish clearly between the professional engineer and the stationary or locomotive engineer, let alone recognize the consulting engineer. Many who seek to engage a consulting engineer have only a vague idea of his function. Most graduates of our engineering colleges have only a meager understanding of the role of the consulting engineer. Even some consulting engineers lack adequate comprehension of their responsibilities and obligations.

This book presents the major factors relating to the practice of consulting engineering. Part One is concerned with external and professional matters: the function of the consulting engineer and his relationships with clients and others. Part Two deals with internal matters: the organization and management of engineering personnel, and the procedures and methods that the consulting engineer employs.

I trust the book will be helpful to both consulting engineers and those who engage them. May it create better understanding, respect, and cooperation between consulting engineers and their clients. May it encourage consulting engineers to render more competent and efficient service to their clients and to hold high the standards of professional conduct. May it inform and interest students and young engineers seeking knowledge of the consulting engineering profession.

Naturally I have drawn heavily on my own experiences as a consulting engineer. However, I believe the subject as presented is applicable to all consulting engineers regardless of the size or type of their practices.

C. MAXWELL STANLEY

Muscatine, Iowa
February, 1961

Contents

vii

The Consulting Engineer and the Client

part one

This part of the book, comprising nine chapters, is concerned with the consulting engineer's professional relationships to his client, to other consultants, and to the public.

The consulting engineer plays an important role in modern economy and technology. He furnishes valuable professional services to individuals, to municipal, state, and national governments, as well as to industrial and commercial organizations.

chapter **1**

The Role of the Consulting Engineer

Contact

From the time he rises and turns the tap for his morning shower until he snaps off the light before going to bed at night, every person who is a part of our industrialized economy benefits in some way from the service of consulting engineers.

The generating stations that provide the electricity for cooking, lighting, and myriad other conveniences he takes for granted were probably designed by consulting engineers. The pure and potable water in his home is very likely supplied by water systems designed by consulting engineers. Many of the highways, streets, roads, and bridges upon which he drives have been planned by consulting engineers. Consultants probably played a part with architects in the design of the office or factory in which he works and the school that his children attend.

It is impossible to imagine contemporary life without the consulting engineers' contribution to such fields as health, sanitation, industrial production, transportation, public works, buildings, utilities, and communications.

The magnitude of the consulting engineering profession in the United States is indicated by the fact that over 7500 firms offer their services in this field. Together they employ some 160,000 people, of whom about half are engineers. Consulting engineering is a rapidly expanding segment of the whole profession of engineering. The number of firms has increased 35 per cent in the last five years and over 100 per cent in the last ten years. Today they employ between 10 and 15 per cent of all engineers in the United States.

But, in spite of the importance and magnitude of the consulting engineer's contribution, relatively few people have direct contact with him or know of his work.

Challenge

Consulting is no field for a person who hesitates to face new problems. The competent engineer who likes variety and enjoys the challenge of ever-changing problems can find a fascinating and satisfying career as a consulting engineer.

Projects handled by consulting engineers include practically every kind of structure, plant, or facility that is planned, designed, constructed, or operated. All the known branches of engineering are covered. The profession demands a high level of engineering competence and ability. As a result, consulting engineers number in their ranks outstanding specialists in many fields of engineering.

There is an added challenge in the area of human relations, since consulting engineers work closely with people both within and without their own organizations. They assume a high professional responsibility in relationships with clients.

The profession demands business and management ability in addition to engineering and professional talent. It offers a unique opportunity for self-employment, a condition seldom found in the engineering profession. It is perhaps the only path open to an engineer with the spirit of an entrepreneur who wishes to become his own boss in full-time engineering work.

All of these factors combine to make consulting engineering a fascinating and exciting career with ever-changing conditions, problems, and contacts.

The Unknown

In spite of its importance, challenge, and fascination, the profession of consulting engineering is little recognized and poorly understood by the public. Actually the man in the street has difficulty understanding the function of any engineer, for he is thoroughly confused by the misapplication of the title "engineer" to many tradesmen, mechanics, and skilled workmen. However, the lack of knowledge of the role of the consulting engineer springs from a more basic fact. Since the consulting engineer usually serves other engineers or organizations rather than the general public, only a few people come in direct contact with him. His services are seldom performed for individuals, as is the case in such better known professions as medicine and law.

What then is a consulting engineer? How does he operate? What does he do? What are the mechanics involved in a successful consulting engineering organization? These are the questions that are discussed in this book from the viewpoint of the consulting engineer and his client. These observations are offered in the hope that they will also be useful to students and others interested in consulting engineering as a profession.

Definition

A consulting engineer is an independent, professional engineer who performs professional engineering services for clients on a fee basis.

Consulting engineers are "independent contractors" in a legal sense, for they own and manage their own businesses and serve their clients on a contract basis. They operate as individual proprietors, partnerships, or corporations. Their staffs vary from a lone secretary in a small operation to several hundred engineers, draftsmen, and other employees in larger firms.

Consulting engineers must be qualified by education and experience to give competent engineering services. They must be registered as professional engineers in the states where they practice. Legitimate consulting engineers have no commercial affiliations with manufacturers, material suppliers, or contractors. They have nothing to sell except service, time, knowledge, and judgment. The compensation of an ethical consulting engineer consists solely of fees paid by his clients for his services.

The clients of consulting engineers are drawn from individuals, industrial and commercial concerns, and municipal, state, and national governments. Any individual or organization needing engineering services is a potential client.

Consulting engineers, who are found in all the principal fields of engineering, offer a range of services as broad and varied as the needs of their clients. Their work varies from brief consultation to the complete planning, design, and supervision of a major project. Normally the consulting firm serves a number of clients simultaneously.

Opportunity

The growing population of the United States, together with the strong trend toward urbanization, industrialization, and mechanization, constantly increases the demand for all types of engineering services. Consequently the role of the consulting engineer is becoming better understood, and the value of his services is increasingly recognized, thus creating a greater demand for his services. To meet this demand, the number and size of consulting engineering firms are expanding, providing more jobs for engineers, technicians, and other employees. As consulting firms grow in size and deal with more complex problems, opportunities for highly skilled and competent engineers increase.

Similarly, there is a continuing opportunity for qualified engineers to become owners of consulting engineering practices, either by advancement within established firms or by the establishment of new firms.

Thus, there are expanding opportunities in the consulting engineering field, both for those seeking employment as engineers and technicians and those wishing to become their own bosses.

Terminology

Here it may be well to define clearly and to limit the meanings of certain terms that are used repeatedly in this book.

The terms "consulting engineer," "consultant," and "engineer" are used interchangeably. They refer to any organization that practices consulting engineering either as an individual proprietorship, a partnership, or a corporation. In general, these terms include staffs and employees of the organizations performing consulting services; in particular, they include the professional engineers of such staffs.

This discussion is not concerned with employees of manufacturing companies or governmental agencies, even though they may be called "consulting engineers" within their own organizations.

Since the function of the consulting engineer is the performance of professional services, our definition excludes construction contractors, sales organizations, and manufacturers. This is true even though such concerns do certain engineering work for prospective purchasers of their services or equipment. A few such organizations purport to offer genuine consulting engineering services, independent and unrelated to their other activities. Only to the extent that such services are truly independent and professional do they come within our definition. Also excluded are such services as laboratory research and management consulting. It is difficult to draw the line between such activities and those that may be clearly classified as consulting engineering. However, the discussions here are confined to the actual practice of consulting engineering.

Emphasis

This book is not concerned with basic engineering theories and practices; much literature on these subjects is already available. It starts with the premise that consulting engineers, being qualified professional engineers, possess sufficient engineering knowledge to be competent in their chosen field.

Accordingly, this discussion is devoted to the professional and management aspects of the consulting practice. The professional aspects emphasized are those fundamental to a proper professional relationship between the consulting engineer and his clients. The management aspects emphasized are those peculiar to consulting engineers that aid in sound engineering practice and that encourage a proper professional attitude.

It is revealing to examine the composition of the consulting engineering profession. Its considerable diversity sets it apart from the better known professions. This diversity creates many of the problems of management and operation encountered by consulting engineers.

chapter **2**

Consulting Engineers

Vital Statistics

Even a casual examination of the engineering profession discloses its tremendous variety as it applies to fields of engineering, clientele, and scope of services. The study of vital statistics relating to size, location, organization, age, clientele, fields of service, and related subjects, which is presented in the following pages, emphasizes further the variety in the consulting engineering profession.

Magnitude

Unfortunately, accurate information is lacking on the number of consulting engineers in the country. No such data are compiled by the Bureau of the Census or other official agency.

The magazine *Consulting Engineer* estimated there were 7500 firms in private practice in the United States in July, 1957.

Based upon 2300 questionnaires returned on a survey late in 1960, the editors determined that the average firm in this country had about 21.2 employees. These estimates would indicate a total employment of consulting firms in the neighborhood of 160,000.

A survey conducted by the magazine *Heating, Piping and Air Conditioning*, published in March, 1959, found 8095 consulting engineering firms in the United States. This survey was based on a tabulation of listings in the classified section of telephone directories.

United States census data indicate that the total number of "engineers" in this country is 528,000. Other data have indicated some 800,000 broadly classified as engineers.

It is estimated that about 75,000 employees of consulting firms would be classified as "engineers." If so, it would seem that between one tenth and one seventh of the "engineers" in the country are so employed. Engineers, as defined by the Bureau of the Census, include many individuals who are not registered professional engineers or who are only partially qualified as engineers. The same is undoubtedly true of the estimate of engineering employees of consulting firms.

The editor of *Consulting Engineer* estimates that some 30,000 registered professional engineers were engaged in consulting engineering as principals or employees in 1960. According to the National Society of Professional Engineers, there were about 227,000 registered professional engineers in the United States in 1958. If these estimates are correct, then one out of seven registered engineers is so engaged.

Location

Consulting engineering firms are scattered throughout the entire nation. This distribution, shown in Table 1, is based on data compiled by *Consulting Engineer*.*

In this and the following tables of this chapter the states included in the various regions are as follows:

---

* These and other data credited to *Consulting Engineer* were published in the issues of January and February, 1957 and January, 1961.

TABLE 1. Location of Firms

Region	Per Cent
East	27
South	17
Midwest	27
Southwest	8
West	21
Total	100

East: Connecticut, Delaware, District of Columbia, Maine, Maryland, Massachusetts, New Hampshire, New Jersey, New York, Pennsylvania, Rhode Island, and Vermont.

South: Alabama, Arkansas, Florida, Georgia, Kentucky, Louisiana, Mississippi, Missouri, North Carolina, South Carolina, Tennessee, Virginia, and West Virginia.

Midwest: Illinois, Indiana, Iowa, Kansas, Michigan, Minnesota, Ohio, and Wisconsin.

Southwest: Arizona, New Mexico, Nevada, Oklahoma, and Texas.

West: California, Colorado, Idaho, Montana, Nebraska, North Dakota, Oregon, South Dakota, Utah, Washington, and Wyoming.

Quite naturally the distribution of engineers closely follows concentrations of population, for such areas have greater need for engineering services. As a result, the majority of consulting engineers are located in the larger cities. However, with improved communication and transportation there is a tendency toward decentralization, and no longer are most firms found in New York, Chicago, and the other large metropolitan centers.

With the rapid growth in the number of consulting engineers, most communities of over 25,000 population now have a consulting engineer who offers at least limited service. In addition, many consulting engineers are found in smaller communities.

Size of Firms

The size of consulting engineering firms varies with the volume and type of service offered. The staff of many firms changes substantially from year to year and even from month to month depending up the work load. Very little statistical data are available on this subject. Obviously, the smallest firm is the one-man operation. On the other hand, a number of firms in the United States have staffs in excess of 500.

Some data on the size of consulting engineering firms are available from the survey of *Consulting Engineer* based on a thousand-unit sample of questionnaires returned in 1956. Its editors, who contend this sample is large enough to make conclusions reliable, believe the geographical distribution of the thousand-unit sample is typical. Their survey indicates that the average size of consulting engineering firms when founded was 4.8; consisting of 1.6 owners, 1.7 engineers, and 1.5 other employees.

The survey further reveals that the average size of consulting engineering firms at present is 31.1; consisting of 2.2 partners or principals, 14.5 engineers, and 14.4 other employees.

Indicative of the fluctuation in size are statistics showing an average maximum size of 50.3 at some time during the firm's history and an average minimum size of 4.1. Thus, at some time in its history, the average firm dropped slightly below its size when founded. On the other hand, at some time it had a staff more than 60 per cent in excess of its present size.

The data of this survey are not truly indicative of the "other" employees because the survey includes nonengineering employees of concerns also offering services outside the consulting field. This situation, found particularly in the eastern states, contributes partially to the differences shown between the East

TABLE 2. Size of Firms

Section	Size When Founded	Present Size	Maximum Size	Minimum Size
East	4.8	39.7	102.5	4.3
South	4.2	21.4	35.9	3.6
Midwest	4.1	21.5	41.8	3.7
Southwest	4.9	18.9	34.7	4.1
West	6.0	44.2	64.8	4.8
Total	4.8	31.1	50.3	4.1

and other sections of the country. Regional averages are shown in Table 2.

The 1960 survey by *Consulting Engineer* with 2315 returns shows an average size of firm of 21.21. This consisted of 2.03 owners who are registered engineers, 2.06 other registered engineers, and 17.12 other employees.

Age of Firms

The consulting profession, which is comparatively young, has enjoyed rapid expansion since World War II. This expansion is due to the greater amount of engineering work available and to a greater acceptance of consulting engineers. The comparative youth of engineering firms in indicated by Table 3, taken from the 1956 survey of *Consulting Engineer*.

TABLE 3. Age of Consulting Firms

Section	Less Than 5 Years, %	5–10 Years, %	10–20 Years, %	Over 20 Years, %
East	31	27	20	22
South	35	31	15	19
Midwest	38	27	12	23
Southwest	35	37	14	14
West	30	37	15	18
Total	33.5	30.5	15.5	20.5

The above data reveal that at present a third of the firms are less than 5 years old, another third have been in business between 5 and 10 years, and only a third are 10 years or older.

These data further reveal that 64 per cent of consulting firms now practicing have come into existence since World War II. Their entire experience has been in the prosperous postwar period. Moreover, only one in five firms now practicing was in existence during the depression years from 1931 to 1936.

Consulting Engineer's 1960 survey found 25.8 per cent of consulting firms were less than 5 years old, 25.8 per cent between 5 and 10 years old, and 48.4 per cent over 10 years old.

Combinations

A sizable number of firms combine consulting engineering with other services. The additional services most frequently offered are contracting and architecture. The extent to which combined services are offered is indicated in Table 4, based on data from *Consulting Engineer*.

TABLE 4. Type of Services Offered in Addition to Engineering

Region	Contracting, %	Architecture, %
East	14	27
South	6	22
Midwest	7	24
Southwest	6	19
West	7	19
Total	9	20

Thus, such combinations are found in substantial numbers throughout the United States. It may be noted, however, that the portion of firms engaged in contracting in the East is twice the average of those in other regions. Similarly, the number engaged in architecture is higher in the East. Most firms that offer contracting services provide a "package" or "turn-key" function by performing both engineering and contracting on a project.

Firms offering both engineering and architectural services are often referred to as architect-engineers and may have both registered architects and registered engineers as principals. In other instances, firms that consist predominantly of engineers may have architectural staffs. The expansion of consulting engineering firms into architecture has been encouraged by various United States governmental departments, which have awarded architect-engineer contracts during and since World War II.

Those firms offering combined engineering and contracting services on turn-key projects are active primarily in the industrial and utility fields. Few operate in the municipal or governmental fields because governmental agencies normally award construction contracts by competitive bidding. Consequently, the engineering design must be finished prior to the awarding of construction contracts, and usually an engineering firm is prohibited from bidding on its own design.

Although contracting and architecture are the functions most usually combined with engineering services, others are sometimes encountered. Consulting services in specialized fields may be offered by manufacturers, for example. Generally this is done through a separate department or division of the company, but not always with a high degree of autonomy.

Occasionally consulting firms attempt to supplement their engineering services by acting as sales representatives for manufacturers. If such a firm is to operate ethically, there must be a complete separation of the two functions. No firm can ethically represent a client as a consulting engineer and simultaneously approach it as a sales representative on behalf of a manufacturer.

Some consulting engineering firms offer services in the management field; still others may be engaged as financial consultants. Since such services are professional in nature, they need not conflict with consulting engineering.

Several of the large "engineering firms" operating principally in the utility and industrial fields offer a broad combination of services, including engineering, architecture, contracting, management and financing. Whether such organizations can, strictly speaking, be classified as consulting engineers is open to question. Nevertheless, they advertise as such and offer services in competition with the true practitioner of consulting engineering.

Fields

The survey conducted by the magazine *Consulting Engineer* explored the extent to which consulting engineering firms practice more than one field of engineering. Current participation in major fields is indicated in Table 5.

As can be seen, most firms practice more than one branch of engineering. It would also seem that, as firms grow older and larger, they seek increased diversification by expanding into the various fields of engineering. Moreover, larger projects normally involve several fields on a single job.

TABLE 5. Fields of Practice

Region	Mechanical, %	Civil, %	Electrical, %	Structural, %	Chemical, %	Other, %
East	62	43	52	43	13	26
South	50	54	49	48	9	24
Midwest	57	54	50	44	10	32
Southwest	58	53	58	45	9	31
West	46	56	45	51	5	22
Total	55	51	50	46	9	27

Table 5 would indicate that the average firm is practicing in 2.4 branches of the profession. It is probable, however, that some of the other branches in the tabulation above are subdivisions of the principal branches; thus the average diversification is less than indicated. Even so, the degree of diversification is substantial.

Clientele

The number of consulting engineers who work exclusively for other engineering firms is very limited. Besides other engineering firms, there are five principal sources from which clients come: private industry, municipal or local governments, state governments, federal governments, and architects. Here again, most firms seek diversification, although a few specialize in one type of client. The degree of diversification is indicated by Table 6, taken from the 1956 survey by *Consulting Engineer*.

These data indicate that the average firm draws clients from

TABLE 6. Type of Client

Region	Private Industry, %	Local Government, %	State Government, %	Federal Government, %	Architects, %
East	89	25	33	29	42
South	81	23	48	29	52
Midwest	91	34	50	20	28
Southwest	87	39	46	23	43
West	89	23	44	41	50
Total	88	28	44	29	42

2.3 of the major sources. In analyzing these data, it should be borne in mind that the percentages represent firms, not dollar volume of business or number of clients.

Insofar as geographical range is concerned, the 1960 *Consulting Engineer* survey indicates that 35 per cent of the firms have a practice limited to one state, whereas 61 per cent operate on a wider geographical base in the United States. It also indicates that 11 per cent of the firms have done some foreign work. However, only 2 per cent of the 1960 workload came from foreign work. Firms tend to start in a narrow area and to widen their operations as they grow. Today, with increasing emphasis and opportunity in foreign fields, there is a distinct trend toward expansion into overseas areas.

Form of Organization

Consulting engineering firms are predominantly sole ownerships or partnerships. There is, however, a distinct trend toward the corporate form of ownership. Since the pros and cons of various forms of ownership are discussed in Chapter 10, we are concerned here only with statistics.

TABLE 7. Type of Ownership as First Organized

Section	Sole Ownership, %	Partnership, %	Corporation, %
East	60	30	10
South	60	33	7
Midwest	65	28	7
Southwest	40	51	9
West	59	26	15
Total	60	30	10

TABLE 8. Type of Ownership at Present

Section	Sole Ownership, %	Partnership, %	Corporation, %
East	50	33	17
South	42	37	21
Midwest	52	28	20
Southwest	43	42	15
West	50	31	19
Total	48	32	20

The survey of *Consulting Engineer* contains information on types of ownership as shown in Tables 7 and 8.

These data indicate the following trends, which are supported by other observations:

1. More than half the firms start as sole ownerships.
2. As firms grow, a sole owner is likely to bring in one or more partners.
3. There is a distinct trend toward increased use of corporate structure.

Regardless of the form of organization, a very high percentage of the ownership of consulting engineering firms rests within the firm. The *Consulting Engineer* survey indicated that in only 3 per cent of the firms in the country was any portion of ownership held by outsiders. The overwhelming portion of ownership is in the hands of registered professional engineers. The same survey, however, indicated that 22 per cent of the firms were partially owned by persons who were unregistered. In most instances this ownership was held by estates of deceased partners or by young engineers not yet registered.

The 1960 survey by *Consulting Engineer* found 55 per cent of consulting firms which were sole ownerships, 25 per cent were partnerships and 20 per cent were corporations.

Age of Principals

The recent rapid expansion in the number of consulting firms has tended to lower the average age of consulting engineers. Nevertheless, only a small percentage of principals are young men; a great number of firms have principals who are in their seventies. Undoubtedly the great majority of principals are between 30 and 60 years of age, and only a few are in their late twenties.

The Start

There is no standard pattern by which engineering firms are started. However, a few general observations indicate the normal pattern.

Only rarely does an engineer become a consulting engineer immediately upon graduation from an engineering college. In nearly all states registration as a professional engineer is required; and some experience is necessary before registration can be completed. Moreover, clients expect experience in their consultants and hesitate to engage an engineer lacking it. Finally, most engineers require a few years to accumulate sufficient capital to permit a venture into consulting practice. As a result, most of those who embark on consulting practices have had a number of years of employment by others. They may have worked with consulting engineering firms, leaving them finally to start their own practice. They may also have worked for various other employers, governmental or private, and, after accumulating experience, decided to enter practice for themselves.

Sometimes an engineer obtains such specialized knowledge and experience in a limited field while working for an employer, that a real demand develops for his services. This demand may lead him to leave his employer and embark upon a consulting practice.

Composite

So varied is the consulting engineering profession that it is difficult to picture a representative composite. An analysis of this chapter shows a sharp contrast with the conditions encountered in the legal or medical professions. In those two fields, practitioners offer their services to the public either as individuals or partnerships, and corporate practice is unknown. Integration with other professions, services, or organizations is rare, and there is less range in the size of organizations.

The great variety in the characteristics of consulting firms creates several sets of conditions and problems within the profession. Obviously the consulting engineering firm with a staff of several hundred encounters different problems from those that confront the lone engineer who practices in a specialized field. This fact must be recognized in considering the relationship of a consulting engineer to his client and as one considers the mechanics of a consulting practice.

Numerous types of service are performed by consulting engineers. This affects many features of the profession, including the size and organization of consulting practices, together with their methods of operation and management.

chapter **3**

Services Performed

Classification

Services performed by consulting engineers may be grouped not only by the branch of engineering, that is, civil, mechanical, and electrical but also by the scope of work. The scope offered depends to some extent upon the type of clients served; it also varies with the size of project handled. All of these classifications and variations are considered below.

Fields of Engineering

Consulting engineers are practicing in every major field of engineering and probably in every principal subdivision of these fields. Actually, there is an unknown number of "engineering fields." Besides the basic branches of electrical, chemical, civil, mechanical, and mining, there are dozens of subdivisions and off-shoots. The directory of engineering organizations, as published in *Who's Who in Engineering*, lists dozens of engineering societies of a technical nature, each dealing with a specific field of engineering.

The number of fields of service is further indicated by an examination of the classified telephone directory for Manhattan in New York City. Six pages of three columns each are required for the listings under the general classification of "Consulting Engineers." Some 600 different firms of engineers come under the following classifications:

Accoustical	Electrical	Metallurgical
Aeronautical	Electronic	Mining
Appraisal	Foundation	Nuclear
Air Conditioning	Gas & Oil	Radio
Architectural	Hydraulic	Refrigerating
Automotive	Industrial	Safety
Chemical	Management	Sanitary
Civil	Marine	Structural
Consulting	Mechanical	Textile

Specialist versus Generalist

Services performed by a given consulting engineer depend not only on his field of engineering but also on his choice between special and general practice.

Some consultants confine their activities to a limited field in which they function as specialists. They tend to become primarily consultants to other engineers or clients having need of highly specialized advice and guidance. Such specialists nearly always work with the engineering staffs of those they serve. Other consulting engineers may elect to specialize in services to a certain type of client rather than perform in a broader range; that is, they may do only municipal work or perhaps only industrial work.

Most firms, however, offer broader service or general practice, although activities may be confined to one or more fields of engineering or one or more types of projects. Such general practitioners are more likely to have several kinds of clients and to perform more varied services. They will also probably be called upon to handle all aspects of engineering projects.

Scope of Services

The range of services performed by consulting engineers is not fully indicated by a discussion of the fields of engineering and types of clients. In addition to the differences above, there is further variety in the scope of work that a consulting engineer may do on a given engagement. It ranges from a simple consultation requiring only a few hours, to the complete engineering of a major and complex project, one that perhaps involves hundreds of man-years of time.

The principal types of services performed by consulting engineers are outlined and discussed in the remainder of this chapter.

Consultation

From this function has come the term consulting engineer. It is the service most generally understood by the public, for it is more similar to the professional services of attorneys and doctors. Consultation occurs when a client avails himself of the expert knowledge and the experience of a consulting engineer. He needs an opinion on some engineering problem or on some procedure, program or project that involves engineering matters. Consultations may be brief or extended. Some involve only a few hours of time, with the client sitting across the desk from the consultant. Other consultations may require considerable travel and a substantial portion of a consultant's time over a period of several months.

Investigation

Most consultations require some study and investigation on the part of the consulting engineer. Such studies and investigations are extremely varied as to type and magnitude. Sometimes they involve only analysis or simple computation that can be undertaken in the office. At other times they require field trips to observe and inspect equipment, apparatus, structures, or projects. Still again, they may involve a review of studies, reports, investigations, or communications prepared by other engineers or by the client's management. The results of studies and investigations are usually submitted in written form, either as a letter or in a more formal report.

Feasibility Reports

Consulting engineers are well known for their feasibility reports. Sometimes such reports are referred to as preliminary, project, or engineering reports. These reports are concerned with determining the feasibility of some project, product, machine, or structure and present the results of surveys, studies, and investigations. They may be concerned only with engineering feasibility but more often also involve the economics of the project.

A feasibility report will usually include such items as purpose of study, requirements and needs of project, alternate solutions, estimated construction costs, estimated annual costs, conclusions, and recommendations regarding engineering and economic feasibility. Feasibility reports are usually submitted in bound form. Sometimes they are typewritten but more often are reproduced by a duplicating process. Frequently they contain drawings and charts graphically supplementing the written text.

Design

Design work probably exceeds in magnitude all other services rendered by consulting engineers.

Engineering design is the process of determining the physical characteristics and dimensions of a machine, structure, device, or project that is to be manufactured or constructed. These characteristics and dimensions are presented graphically on engineering drawings, commonly referred to as blueprints by the layman. Such drawings, or plans, are supplemented by written material called specifications. Plans and specifications are used to direct the manufacturer or the contractor in the fabrication or construction of the device, machine, structure, or project.

However, design work is not limited to mathematical computations and to the preparation of plans and specifications. It first requires study, investigation, and research. Often it entails careful comparison of available materials, machinery, and equipment. It may include analysis made on electronic computers or network analyzers. Design usually involves comparison of alternates, both from economic and engineering points of view, to determine the most suitable selection. Frequently the design process includes the preparation of detailed lists of materials, called bills of materials. These are used to procure or purchase the parts and materials required to manufacture or construct the work.

Procurement

Once the designs are completed in the form of suitable plans and specifications, the consulting engineer often assists the client in selection and purchase of materials or in the award of construction contracts.

Procurement or purchase usually involves the receipt of a proposal from one or more suppliers of the required equipment or material. Selection may be made on a competitive or a negotiated basis. The engineer may assist the client in evaluating proposals and recommending desirable selection to the client.

On construction projects, particularly for government organizations, construction contracts are usually awarded as a result of competitive bidding. On such projects the engineer normally prepares contract documents in addition to plans and specifications, usually in cooperation with the client's legal advisor. When the plans, specifications and contract documents are ready, bids are solicited from contractors or manufacturers through public notices issued in accordance with legal requirements. When proposals are received they are opened, publicly, as a rule, read and tabulated by the consulting engineer who then makes his recommendations to the client. Once the purchasing process or contract award is completed, the consulting engineer often assists the client in the preparation or review of the contract or the purchase order used to acknowledge the commitment.

Construction Supervision

Normally the consulting engineer who has designed a project will be responsible for supervising construction. This activity consists of two parts—general supervision and resident supervision.

The following services come under the heading of general supervision:

Periodic visits to the site of the work.
Consultation with the owner.
Interpretation of plans and specifications.
Checking shop drawings and data.
Processing and certification of contractor's payment estimates.
Preparation of amendments to contractor's contract.
Final inspection of project.
Preparation of "as constructed" drawings.

Under the heading of resident supervision, the consulting engineer sends a representative, known as a resident engineer to the site of the project. This individual is responsible for detailed supervision and inspection to ensure that the project is constructed according to the plans, specifications, and contract documents. In addition, he also coordinates and expedites the work of the contractors. Such a resident engineer may be assisted by one or more inspectors if the size of the project warrants it.

As a part of construction supervision, the consulting engineer may send inspectors or engineers to the plants or shops where equipment or materials are being manufactured for use on the project. Their mission is to inspect the apparatus or materials in process of manufacture, to observe its performance in test, and to assure its compliance with the plans, specifications, and contract conditions.

Testing

Frequently consulting engineers are responsible for testing equipment, apparatus, or complete plants. Such testing may be necessary to insure satisfactory operation or to determine precisely the efficiency or operating characteristics of the equipment. In any case, the purpose is to make sure that performance matches the guarantees that may have been made regarding operation or efficiency. A certain amount of testing is normally associated with the engineering on new equipment, apparatus, or projects. Sometimes consulting engineers also test existing plants or equipment.

Production Engineering

A huge field of engineering related to the operation of manufacturing or processing plants, as contrasted with their design and construction, exists, and frequently consulting engineers participate in this work, cooperating, in such cases, with the regular staff of the plant. The work may include process engineering, production engineering, methods, tooling, time study, quality control, and other related functions.

Valuations and Rate Studies

Some consulting engineers practice in the fields of valuation and rate making. A valuation is a determination of the worth or value of a plant or property. It is used to determine rates for utility services and to establish proper value for insurance purposes or for purchase or sale. Since valuations are a prerequisite to most rate-making procedures, consulting engineers may practice both fields. Determination of rates brings the engineer into a field that involves many legal proceedings and often requires appearances before public service commissions or courts.

Surveys

Consulting engineers, usually individuals or smaller firms, may also make land or property surveys unrelated to a specific project. Nearly all engineering engagements that include the design of projects involve some topographic or site surveying.

Operation

Engineers may render a supervisory operating service in connection with newly completed projects. This service may continue until the client's personnel have become adequately familiar with the project. On the other hand, a few consulting engineers regularly offer services related to the operation of plants or facilities. These are usually of an advisory or consulting nature and involve an analysis of records of performance supplied by the client.

Court Work

Often consulting engineers are requested to function as expert witnesses in court proceedings and to advise clients and attorneys on engineering matters involved in legal procedures. A limited number of consulting engineers specialize in court work, particularly in the fields of valuation and rate making.

Complete Services

Quite frequently consulting engineers who concern themselves with projects that are constructed render complete project engineering including:

Feasibility report.
Design.
Procurement or award of construction contracts.
General supervision of construction.
Resident supervision.

This list is a normal package of engineering services on engagements that result in the construction of plants, works, or projects.

Other Services

The services outlined above cover the principal ones performed by consulting engineers, but the list is by no means complete. Many consulting engineers perform other services of a specialized or limited nature. For example, there are engineers who engage in patent work, often working closely with attorneys. Other consulting engineers are engaged in work related to financing and accounting, and serve financial institutions or similar organizations. Still other consultants handle research problems and aspects of management or operations of one type or another. The compilation of a complete list of services would be a formidable task and would serve no useful purpose. The listings given here adequately indicate the range of services performed by consulting engineers.

The consulting engineer has an intimate professional relationship with his clients that requires mutual understanding, cooperation, and respect.

chapter 4

The Consultant and His Clients

Who Is the Client?

A complete list of all the types of organizations, industries, and institutions using the services of consulting engineers would be a lengthy one. Preparation of such a list might be interesting but it is more pertinent to outline the few basic categories in which clients are found. Any individual or organization needing engineering services is a prospective client. Thus, clients of consulting engineers come from a wide field: governmental bodies, industries, businesses, commercial organizations, and individuals.

Governmental clients include subdivisions, departments or agencies of the national government, states, counties, and municipalities. They also include independent and quasi-public authorities, districts and commissions authorized by national, state, county, or municipal law.

In the industrial, business, and commercial fields, clients are usually corporations, although they may be partnerships and individuals. Clients are found in every field of endeavor that requires plants, machinery, structures, or facilities that must be planned, designed, constructed, operated, and maintained.

Frequently consulting engineers have other engineers or architects as their clients. In such cases they function in an advisory capacity or handle some portion of the engineering on a project for which the other party has the primary responsibility.

Individuals seldom require engineering services except in connection with a business or commercial activity. For this reason consulting engineers seldom work with individuals but with a group or an organization. Thus, although the consulting engineer is practicing a profession, his is less personal than the better known professions.

Why Clients Use Engineers

Clients may use the services of consulting engineers for any one or a combination of reasons, for example:

1. The client has no engineering staff.

2. The client's engineering staff is inadequate in size to handle the required engineering work load.

3. The client's engineering staff lacks the competence to handle a given engineering problem.

4. The client desires to have engineering responsibility shared with or carried by an outside consultant.

5. The employment of consulting engineers avoids distraction from the client's normal operations.

6. Legal statutes require the engagement of an independent consulting engineer.

7. The use of a consulting engineer is more economical.

Each of these reasons is analyzed and discussed in the following pages.

No Staff

If an individual or organization is confronted with an engineering problem and has no engineering staff, there are only two choices: engage a consulting engineer or build the required engineering staff. Many small industries and municipalities are in this situation because their normal operations do not justify an engineering department. The occasional engineering problem or project that confronts them can best be handled by the retention of a consulting engineer for several reasons. It saves time because a consultant can be engaged and put to work quickly, whereas the establishment and recruitment of an engineering staff, or even one engineer, takes time. It assures engineering competence and experience, whereas the ability of a new staff cannot be determined until it has functioned for a period. It avoids the complications and difficulties of putting together a new organization. For all of these reasons, it normally is far more economical to use a consulting engineer, particularly in connection with an occasional nonrecurring engineering need.

Overflow

An organization with an engineering staff may be confronted with a work load beyond its capacity. Often an industry with a plant engineering department staffed to handle maintenance and minor improvements undertakes major plant expansions. Similarly, a municipality, with a city engineering department handling routine sewer and street extensions, may undertake a large program of trunk sewers and a sewage treatment plant. In such cases, the prospective client must either expand his staff or engage consulting engineers. The advantages of using a consultant are the saving of time and expense, the avoidance of the difficult problems of recruitment, and the assurance of engineering competence. Furthermore, the engagement of a consultant allows the client's engineering department to keep abreast of its normal and continuing assignments and responsibilities.

Competence

Often a client with an engineering department encounters requirements for engineering services that are beyond the competence of its organization. For instance, it may undertake a different type of work or face an extremely complicated or difficult engineering problem. Under such circumstances, the client may engage a consulting engineer to take advantage of his experience and expertness with that particular problem. The consultant may either handle the entire problem or project, or he may simply confer and advise with the client's engineering staff. In either case, the client is assured of the skill, experience, and competence that are needed, and he obtains these more quickly and at less cost than he would by expanding his staff.

Responsibility

In situations involving difficult or controversial matters, the client may engage a consulting engineer to carry or share the responsibility for a particular project or decision. The client may do this even though his own engineering staff is capable of handling the matter. In other situations a client may retain a consulting engineer to review engineering work that has been done by the client's staff, in order to have him share the responsibility. The engagement of an independent consulting engineer in such situations gives assurance to the client's management and strengthens its position with investors, commissions, the public, or others who may be concerned with the action taken.

Distraction

One of the soundest reasons for the engagement of consulting engineers is to avoid distracting the client from his normal functions and operations. If the client's staff, both managerial and engineering, is of the proper size for normal operations, it will be inadequate to handle additional major projects or programs. By engaging consulting engineers and by shifting substantial responsibility to them, the client lessens the work load carried by his management and engineering staffs. Thus, the distraction from the normal problems of production, operation, and maintenance is kept to the minimum.

Legality

The laws of many states require governmental or quasi-public agencies to employ independent consulting engineers for certain functions or projects, for instance, in Iowa drainage and levee districts, organized under the provisions of the state law, are required to employ engineers to prepare certain reports and designs for improvements. Under these and similar conditions, the client has no alternative but to engage consulting engineers.

Economy

Economy is a usual by-product of the engagement of consulting engineers. Sometimes, however, it is a primary reason for such action. A consulting engineer's services are usually more economical because of:

1. The availability of an experienced, competent staff ready to undertake the engagement on short notice.
2. The experience and skill in organizing and administering a project or an engagement.
3. The background of knowledge derived from experience with similar projects for other clients.
4. The ability to utilize staff only as needed and then assign them to work for other clients.

The costs of engineering done by a consultant are readily determined from the fees paid him. Comparative costs of engineering work undertaken by a client's own staff are more difficult to obtain, and these costs are often surprisingly high if proper accounting is made to include all overhead and hidden costs. Sometimes clients unfairly compare the estimated direct salary cost of their own engineers against fees of consulting engineers. In order to reach a fair comparison, a full and proper allowance must be made for all of the overhead costs involved in maintaining an engineering department. (See Chapters 17 and 18.) When this is done, the engagement of a consulting engineer is usually found to reduce engineering costs.

Professional Relationship

The relationship between the consulting engineer and his client is a professional one, not unlike that found in the medical and legal professions, but naturally there are significant differences. The consulting engineer is most often engaged by an organization rather than an individual. Generally his work is concerned with contractors and manufacturers who are building or supplying equipment and materials for the client's projects.

Historically the professional man has been identified by two criteria: he must be highly competent as a result of his education, preparation, and experience, and he must assume a responsibility to his client or patient and to society that transcends personal gain. The nature of this professional attitude and responsibility is examined more fully in Chapter 9.

Ethics Toward Clients

Many codes and canons of ethics have been prepared to govern the engineer's professional conduct. The fundamental principle of ethics as they affect the client is that the consulting engineer will render faithful, professional service to him and honestly represent his interests.

Several of the well accepted precepts falling within this relation between the consulting engineer and his client are stated below:

1. The consulting engineer shall protect his client's interests in every legitimate way as a faithful agent or trustee. Acting for his client in a professional capacity, he has a responsibility far beyond the legal obligations expressed in the engineering service contract and of a higher level than that involved in a normal business transaction.

2. The consulting engineer will not divulge any confidential information obtained from a client. He may use information sup-

plied by his client which is not common knowledge or public property. Such information will be kept absolutely confidential and considered the property of the client.

3. The consulting engineer will inform the client of any business connections, interests or affiliations that he has which might influence his judgment or impair the quality of services he can render.

4. The consulting engineer will recognize any limitations in the ability and experience of his organization and, when advisable, consult, retain, or cooperate with experts and specialists.

5. The consulting engineer will not accept any trade commissions, discounts, allowances, or other indirect profit or consideration in connection with any work he performs for a client. His sole source of income on any engagement will come from the fees to be paid by the client.

6. The consulting engineer will present clearly to the client the consequences to be expected from proposed deviations if his engineering judgment is overruled in areas where he is responsible for technical adequacy of engineering work.

7. The consulting engineer will act with fairness and justice in his quasi-judicial position as arbiter between the client and the contractor.

Client Ethics

No code has even been written by clients outlining ethical practices toward the consulting engineer. Were such a code to be prepared, it would surely contain those items essential to mutual respect and to a sound professional relationship. A few suggestions for such a code follow:

1. The client will select engineers on the basis of merit in accordance with accepted practices.

2. The client will pay adequate and equitable fees to the consulting engineer for his services.

3. The client will accept the consulting engineer as a professional advisor and extend to him the respect and confidence warranted in such a relationship.

4. The client will accept responsibility for the consequences to be expected from proposed deviations from the consulting engineer's recommendations.

5. The client will promptly furnish the consulting engineer the information and data that he requires, and will promptly review and approve the information and recommendations submitted by the consulting engineer.

6. The client will give credit to the consulting engineer for his part in the engineering work.

7. The client will respect the consulting engineer's position with respect to contractors and material suppliers on construction work and will not bypass him by going directly to the contractor.

Understanding

If the consulting engineer is to discharge his responsibility properly and to give full service value, there must be good working relations with the client. Yet, the establishment of proper working relations between the consulting engineer and the client is not a simple matter. It imposes a responsibility that both parties must fully assume.

The first point in a good relationship is understanding. The client must understand the functions and role of the consulting engineer. If he does not, how can be cooperate fully with the consultant? Accordingly, one of the engineer's first responsibilities is to make certain that his client does understand the function he will play and the method in which he will operate. Such understanding requires education, which may be partially accomplished during negotiations and preparation of the engineering service contract. Beyond this, however, the consulting engineer must secure from the client sufficient time to permit a detailed outline of the engineer's procedures and methods of operation.

It may seem elementary to emphasize this step, but its importance has been driven home to me by experience with uninformed clients. It is amazing how often the consultant finds client's representatives who really do not fully understand the role of the consultant. A general understanding of the function of consulting engineers needs to be supplemented by specific knowledge of his procedures and methods. Sound education on these points will avoid many difficulties.

An equally important area of understanding concerns the client's needs and desires. Too often a consulting engineer accepts an engagement, signs a contract, and rushes to work without fully understanding the client's problems. Here, too, the consultant should take the lead in making certain that there is a full meeting of minds. This is particularly true when extensions or modifications of an existing facility are involved or when the client's staff has already studied and planned the project.

The Three C's

Besides understanding, three other major factors are involved in good client relations—communication, cooperation, and confidence.

It is incumbent upon both the consultant and the client to see that proper channels of communication are utilized. First, this requires agreement as to the addressees of letters, communications and reports, and the frequency of written reports or personal contacts. Next, it requires the use of these channels of communication once they are established. There is no better way to avoid misunderstanding and dispute during an engineering engagement than to assure good communication.

The need for cooperation is more evident but equally important. Cooperation extends to such things as exchange of information, mutual assistance in obtaining data, prompt approval of materials submitted by either party, and prompt handling of correspondence. But beyond such details cooperation requires mutual teamwork of the client's and the consultant's organizations, both working to achieve a single objective.

Confidence is probably the most important ingredient in satisfactory client-consultant relations. Unless the client has confidence in the competence and integrity of the engineer, he seldom allows him to function as he should. And, on the other hand, the consultant must have confidence in the client and those who represent him. If this is not the case, he will perhaps seek to make a record and protect himself rather than give primary attention to the performance of professional services. I know, from personal experience, that lack of confidence not only makes the task more difficult for the engineer but actually impairs the quality of service given the client.

A client should never engage an engineer unless he has confidence in him. Occasionally our firm has been selected by a client, not because of confidence in any engineer but because of a legal or other compulsion. Fortunately, in many of these situations, mutual confidence developed as we worked together and became acquainted. Unfortunately, in cases where confidence did not develop, neither party obtained satisfaction.

Engineering fees are determined by several methods. The level of compensation and the form of the fee structure are each important in establishing proper financial arrangements between the client and the consulting engineer.

chapter **5**

Fees for Services

Compensation

The sums of money paid to consulting engineers for professional services are called fees. The expected compensation is usually determined in advance by agreement between the client and the consultant. A number of alternate methods are available for computing fees. The particular one selected depends on the nature and magnitude of the services to be performed and on the preferences and prejudices of both the client and the consultant.

Methods

The methods of fee determination that have found widest acceptance among consulting engineers are:

1. A lump sum or fixed fee for specified services.
2. Time charges, based on man-hours, man-days or man-months of time expended in rendering the service.

3. Direct salary cost or direct payroll cost plus a percentage thereof.

4. A percentage of the cost of construction of a project.

5. Cost plus a percentage thereof or plus a fixed fee.

6. A retainer.

Often combinations of these six basic methods are used. As a rule, the engineer is reimbursed for specified out-of-pocket expenses in addition to the basic fee. Sometimes a limit or ceiling is placed upon the fee to be earned.

Level of Fees

If a consulting engineer is to perform competently, he must have adequate compensation to cover costs and allow a reasonable profit. On the other hand, the client is entitled to a fee that fairly represents the value of service performed and that does not result in an exorbitant profit to the consulting engineer. But, it is perhaps even more important that the fee be adequate enough to assure high quality professional service. The client is the loser if this is not so; for, if a consulting engineer's fee is inadequate, his only alternatives are to slight the work or to lose money on the engagement. A high caliber professional performance requires adequate time and the attention of competent engineers and staff. If fees are too low, the time spent on the work will be insufficient to permit the best performance. If a job is slighted because of inadequate fee, the inevitable result is second-rate engineering. The client will lose because engineering decisions will not be the best and the function, quality, or the cost of the project will be adversely affected.

Low engineering fees, therefore, are a false economy. This is particularly so because the fee for adequate engineering is quite modest compared to the value of the service. For instance, the total fee to a consulting engineer for complete engineering services on a moderate-sized construction project is less than the profit that the construction contractor normally includes in his bid. To use another comparison, the cost of engineering services on such a project is in the range of the interest cost for one and two years, which the owner pays for the money he uses to finance the project. The payment of fees adequate to assure complete and competent engineering is a prudent and economical investment for the client.

Lump Sum

The lump sum or fixed fee for specified services allows both the client and the consulting engineer to know in advance the exact amount of the compensation, thus avoiding complicated invoicing and accounting. In setting a fixed fee, the consulting engineer must include direct salaries and expenses, indirect costs or overhead, and a reasonable profit. The prudent engineer will also include some allowance for contingencies. A fixed fee can be used intelligently only where the scope of the work is clearly defined. It should not be used for engineering services on a construction project unless preliminary investigations have accurately determined the nature and cost of the project. It is difficult to set in advance for studies, reports, or investigations when the magnitude cannot be accurately anticipated.

In spite of these difficulties, lump sum compensation is frequently used for studies, reports and investigations, and, occasionally, for engineering services on projects involving construction. It is also used as a billing method for engagements without prior arrangement as to fee. One such example is a consultation involving specialized knowledge, where the value of the service has no direct relation to the time expended by the consultant.

Time Charge

The simplest means of determining compensation for consulting engineers is the use of time charges, usually by the day or the hour. This is the familiar "per diem" method of compensation, which has wide acceptance.

The consulting engineer sets daily or hourly rates of charges for the various members of his staff. These time rates include not only direct salary but proper allowance for indirect costs or overhead and profit. Various professional engineering societies usually suggest time rates two or three times the hourly or daily salary of professional personnel. Higher multipliers are recommended for engagements involving expert testimony or requiring professional opinion based upon wide experience.

The selection of the multiplier to be applied to daily salaries to compute per diem rates is a matter of judgment. It must be sufficient to cover all indirect costs as well as a reasonable profit. Some of the factors involved in determining the multiplier are discussed in this chapter under the heading "Level of Fees" and also in Chapter 18.

Per diem rates are usually established for several categories of professional and technician personnel, and sometimes for service personnel. Naturally the level of per diem rates will vary with experience, salary, and competence of the staff, and with the consulting engineer's policy on markup to cover overhead costs and profits.

The top rate for partners and officers probably ranges from a low of $75 to a high of several hundred dollars per day. The following schedule shows relative per diem rates for an assumed firm:

Partner	$150.00 per day
Principal engineer	125.00
Senior engineer	100.00
Engineer	75.00
Junior engineer	50.00
Senior draftsman	60.00
Draftsman	50.00
Junior draftsman	40.00

Higher rates may be listed for expert advice and other work involving an extremely high level of professional skill and knowledge. Hourly rates may be determined by dividing per diem rates by 6 to 8, the hours in a normal work day.

Monthly charges, rather than daily or hourly ones, may be used where a given individual is continually assigned to the work of a single client, such as resident engineers and inspectors on construction projects. Monthly rates, too, must include direct salary, indirect costs, and profit. Often the multiplier for these factors is less for field personnel than for office personnel.

When straight time charges are used, certain out-of-pocket

expenses are normally reimbursable; these are billed to the client at cost. Such items are travel and living expenses for staff away from the home office, identifiable telephone and telegraph charges, costs of blueprinting and duplication, costs of laboratory tests, and other special services.

Time charges, which are easily understood and completely flexible, are well suited to situations where the scope of work cannot be accurately determined in advance. However, the use of time charges has certain disadvantages. Detailed and accurate accounting must be maintained and billings to the client must be fully supported by summaries of time and cost. This system is subject to abuse if work is not carefully controlled. Even where abuse does not occur, the client may suspect the consulting engineer of encouraging or permitting excessive time use. Time charges may also be objectionable to some clients because the level of the per diem fee seems excessive. In spite of these disadvantages, time charges are without a doubt the fairest and most equitable method of determining fees and are, therefore, widely used.

Sometimes a ceiling is applied to set a limiting cost for services. This may be desirable, for it assures the client that fees on an authorized engagement will not go beyond a predetermined amount.

In passing, I would like to note that no type of compensation for engineering services can be free of criticism from a suspicious client. In all types, except possibly the fixed fee, either the time expended, the reimbursable costs, or the construction cost of the project are under the control of the engineer. If mutual confidence does not exist, the consulting engineer can expect controversy over the fee.

Direct Salary Plus a Percentage

Another type of fee used is direct salary plus a percentage thereof and plus specified reimbursable expenses. Such a fee may be stated as follows: (a) direct salary of staff for the time employed on the project, plus (b) a stated percentage thereof to cover indirect costs and profit, and plus (c) reimbursable out-of-pocket direct expenses including travel and living costs for personnel while away from the home office, telephone and telegraph expenses, and costs of blueprinting and reproduction.

The method is similar to a straight time charge on either a per diem or hourly basis as the fee is directly proportional to the hours of service of the engineer's staff. The percentage applied to direct salary must cover all items of an engineer's indirect cost or overhead, together with profit. Factors entering into the selection of the percentage are similar to those previously discussed under "Time Charges."

This method has the same advantages and the same disadvantages as time charges. However, it possesses the additional advantage of expressing the fee as actual salaries rather than as higher hourly or per diem rates. This method of compensation is particularly attractive to industry, which is used to thinking in terms of payroll plus a burden or overhead. A variation of this method uses direct payroll cost as a base. Direct payroll cost is usually defined as direct salary plus those costs intimately related to direct salary, such as payroll taxes and insurance. In determining direct salary, several approaches are used. One method is to divide total annual salary by 52 times the number of hours worked per week. With this method, the cost of vacations, sick leaves, and holidays is considered as an "indirect" salary. An alternate method is to divide the annual salary by the number of regularly scheduled hours of work per year, excluding vacation, sick leave, and holidays. If the first method is followed, then the percentage added to cover indirect costs and profit must be larger than that of the second method in order to provide the same margin of profit to the consulting engineer.

Most recommended fee schedules suggest that the multiplier on direct salary be not less than 2.0, and some schedules suggest situations under which it may be much higher (as much as 3.5). Practice varies but factors seem to range generally between 2.0 and 2.5, depending upon indirect costs and other factors that are discussed more fully in Chapter 18.

Percentage of Construction Cost

Fees expressed as a percentage of construction cost are normal for complete engineering services on construction projects. Such fees are stated in different ways. One example, taken from the recommended fee schedule of The Ohio Society of Professional Engineers, Schedule I (for complex projects), suggests declining percentages, as shown in Table 9.

TABLE 9. Fees as Per Cent of Cost

Cost of Construction, in Dollars	Basic Minimum Fee, %
0–50,000	Per diem only
50,000–150,000	10.0
150,000–250,000	8.5
250,000–500,000	7.5
500,000–750,000	6.5
750,000–1,000,000	6.0
1,000,000–5,000,000	5.75
5,000,000–10,000,000	5.5
10,000,000 and over	5.25

Another, or block form, as illustrated by the schedule taken from the current recommendations of the Iowa Engineering Society and the Iowa Association of Consulting Engineers, Schedule A (for complex projects), is shown in Table 10.

TABLE 10. Block Type Per Cent Fees

Step, in Dollars		Basic Minimum Fee, %
First	25,000 (of construction cost)	12.0
Next	75,000	9.0
Next	300,000	7.5
Next	300,000	6.0
Next	800,000	5.5
Over	1,500,000	5.0

A curve may be used to present a fee schedule. Figure 1 is such a curve expressing median fees for professional services as presented in American Society of Civil Engineers, Manual 38. All of these methods establish a percentage that decreases with the size of project. This fact is demonstrated by Figure 2, which charts several currently recommended fee schedules. The selected percentage fee is applied to the construction cost, which

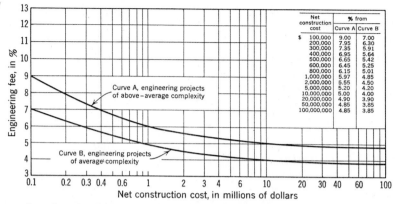

Median fees for fourteen items of professional services as a percentage of net construction cost. The fees are to be adjusted to suit special conditions.

Figure 1. Chart of suggested fees. *Private Practice of Civil Engineering,* ASCE Manual 38.

is usually defined as the total cost of the project exclusive of costs of land, right-of-way, legal and engineering fees, and the owner's overhead and financing costs.

The services normally provided in a percentage of construction cost fee, as shown in recommended schedules, include: preliminary report, design, preparation of plans and specifications, and general supervision of construction.

The following services are seldom included in a percentage of construction cost fee: resident supervision and inspection, revision of plans and specifications after approval, computation of special assessments, property surveys, cost of laboratory tests on materials, and cost of soil borings. Other exceptions are sometimes listed.

Where the scope of services to be provided is other than normal, suitable adjustment must be made in the percentage rate. Sometimes a ceiling is set when using a percentage of the construction cost fee schedule. For instance, some municipalities cannot legally execute a contract without a stated maximum commitment. Frequently certain costs will be reimbursable in connection with a percentage of construction cost contract; for example, travel of the engineer other than between his office and the site of the project.

There is considerable variation in recommendations as to the division of the percentage of construction fee among major phases of the work. Suggested ranges are as follows:

Preliminary report	10 to 20%
Design, including plans and specifications	45 to 65%
General supervision of construction	20 to 40%

The proper division depends on the nature and size of the project.

A percentage of construction cost fee must include all direct and indirect costs and profit for the engineer. Years of experience with percentage of construction cost fees have demonstrated the merit of this method of compensation. If a proper schedule

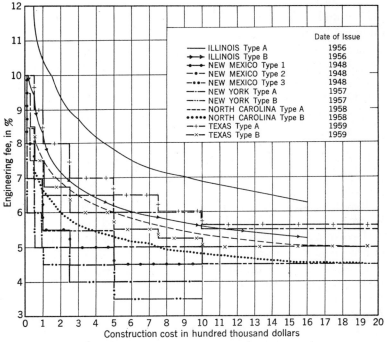

Figure 2. Fee schedules per cent of construction cost.

of percentages is used, it provides fees generally proportional to the engineer's cost of service. Furthermore, it allows flexibility so that, if the scope of the project is varied, the compensation of the engineer is adjusted. This method has the additional advantage that the fee can be easily determined once the construction cost is known. Such a fee includes all engineering cost to the client and there are no special problems of accounting, invoicing, or auditing.

However, the percentage of construction method of computing fees has two major disadvantages. One results from the fact that the fee varies with the construction cost. Therefore, an engineer doing an outstanding job of design whereby construction costs are reduced automatically cuts his fee. On the other hand, a poor job of design or administration by the consulting engineer may increase his fee along with construction cost. The other disadvantage is that the percentage schedules are of necessity arbitrary and average and thus do not take into account all conditions affecting engineering costs. Because of these weaknesses there is a growing tendency to substitute other methods of compensation where there is mutual confidence and respect between the client and the consulting engineer. This tendency should be encouraged.

Cost Plus

With this method of compensation, the consulting engineer is reimbursed for certain defined costs plus either a percentage thereof or a fixed fee.

The definition of reimbursable costs varies greatly. In general, it covers payroll cost, including all direct and indirect salaries and costs associated with such payroll. It also normally covers out-of-pocket living and travel expense, identifiable communication expense, costs of authorized outside services, identifiable costs of supplies and duplication. Also included are such indirect costs as provision of office and working space, rental or depreciation allowances on furniture, equipment, and instruments, and state and federal taxes. It may include certain indirect staff, administration, accounting, and similar functions, although the payroll of owners and top executives is often excluded.

The "plus" must cover the profit to the consulting engineer and any salaries, overheads, and expenses that are excluded from reimbursement.

When the additive fee is expressed as a per cent of reimbursables, the range is often from 10 to 15 per cent. Because of a stigma attached to this type of cost plus, the more normal practice today is to use a fixed fee. In either case, the "plus" is established by negotiation with the client after the definition of reimbursable items has been written.

Cost plus contracts require extreme care in accounting to record expenditures of time and money properly and to document them to assure reimbursement. Advance approval is often required on size of staff, salaries, and other important cost factors. Any change in scope of work on a cost plus, fixed fee contract requires a renegotiation of the fixed fee. In spite of administrative difficulties, cost plus contracts are frequently used. Many agencies of the United States government use them on projects where the scope is poorly defined. They are also used on crash-type programs, such as overseas military bases where engineering costs cannot be estimated in advance.

Retainer

The use of a retainer, common to the legal profession, has some acceptance in consulting engineering. A retainer is a fixed payment to the engineer, on a lump sum monthly or annual basis. In return for this payment by the client, the engineer is ready to serve the client upon call and perhaps to render certain periodic services.

Retainers are most common where the client has a continuing need of engineering advice. The usual retainer agreement provides that if services are required above a stated level, the engineer will receive additional compensation on a stated basis. Sometimes the retainer is simply an advance payment of fees to be offset against actual billings when services are performed. A retainer may be combined with other types of fees. Determination of the size of retainer fees is a matter of judgment based upon the expected need and value of services.

Reimbursable Expense

Any of the various types of fees discussed above may be supplemented by a provision for reimbursement to the consulting engineer of certain out-of-pocket expenses or additive services. As has been indicated, such out-of-pocket costs as travel and living expenses for staff away from the home office, identifiable communication and printing or duplicating costs are frequently reimbursable. In addition, the contract may provide compensation to the engineer for services added to the basic scope of services. These may be billed at normal per diem rates or on a cost plus basis. In all cases of reimbursable items, it is important to have the definition and basis of reimbursement clearly stated.

Ceiling

Occasionally a ceiling or maximum fee limit is stated in an engineering service contract. Such a limit may be used with any of the types of fee structure except lump sum.

A ceiling protects a client by establishing the maximum financial commitment in a given engineering engagement. This commitment may be necessary for legal requirements or may be desired by the client for other reasons. When a ceiling is established, it should normally be somewhat above the estimated fee, perhaps 5 to 15 per cent, to cover contingencies and unexpected items. Ceilings may be used also to provide a control on engineering engagements where the scope is not determinable in advance. Such a ceiling may state that the engineer will not invoice beyond a given amount for services without further approval of the client. If the execution of the work requires fees beyond the ceiling, a new and higher ceiling is agreed upon by the client and the engineer.

Selection of Fee Type

The type of fee to be used for a given engineering engagement is a matter of negotiation between the client and the consulting engineer. Ordinarily the consulting engineer will propose the type of fee he feels is best suited to the particular engagement. In reaching this decision, he must consider the preferences or prejudices of a client, and also any legal requirements applying to the client.

Obviously there are many possible combinations of the several types of fees. It is common to use a percentage of construction cost fee on a project plus a time charge for resident supervision. It is also common to use a lump sum fee for a preliminary report with the amount applicable upon a percentage of construction cost fee for complete engineering services. Other combinations will suggest themselves as situations arise.

Once agreement on fees is reached, it should be stated in a suitable contract (see Chapter 6). This contract should include the conditions affecting time and method of fee payment.

Recommended Fees

The normal guides to equitable fees are the numerous fee schedules published and recommended by various professional engineering societies. Such schedules are generally concerned with fees determined by the time charge method or by the per cent of construction cost method.

Such publications often contain alternate schedules applicable to varying conditions. Frequently they suggest two alternate schedules for determining per cent of construction cost fees on projects of varying complexity. Sometimes they suggest alternate multipliers for application to direct salary or payroll cost, to determine time charges fees. Such multipliers may vary with the magnitude or the complexity of the engagement.

There are substantial differences among such recommended fees. Some of this variation reflects differences in costs of operation in various areas. Part of it comes from the particular experience and viewpoints of those who have compiled the recommendations. Most such recommendations represent a compromise of viewpoints on the part of those who participated in their formation. Also, a few schedules are badly out-of-date and need revision to reflect present day costs. In spite of such variations, these various fee recommendations are helpful guides in determining fees. Every consulting engineer should familiarize himself with suggested fee schedules applicable to the areas in which he is operating and to the types of work he is doing.

Departure from Recommended Fees

The various recommended fee schedules should be regarded as guides, not as rigid requirements.* Each consulting engineer should supplement them with his experience and judgment. Situations often occur where a departure from recommended schedules is appropriate. Many recommended fee schedules give undue weight to certain types of projects and result in fees inadequate for other types. In such cases they are of little help even as a general guide.

Fees higher than recommended schedules are warranted where problems are particularly complex or where clients require special or additional services. Projects involving rehabilitation usually warrant higher fees. Some consulting firms give additional services that call for greater compensation. Other firms, whose services are greatly in demand, may customarily demand higher fees. On the other hand, there are situations that justify fees lower than suggested schedules. Projects that are extremely simple or that involve a great deal of repetition may call for a slight downward adjustment of fee. If the client performs portions of the engineering work or agrees to eliminate portions of it, an adjustment is obviously called for.

However, the recommended fee schedules are the best available information to guide consulting engineers and their clients in establishing proper fees. They should be used for this purpose until a consulting engineer's own experience reveals needed departures or until the composite experience of engineers expresses itself in better schedules.

If consulting engineers are selected on the basis of their qualifications in the method suggested in Chapter 7, the determination of fees is then a matter of subsequent negotiation

---

* One school of thought among engineers contends that fee schedules adopted by engineering societies be considered mandatory and that departure from such schedules be considered a violation of ethics. This position is challenged by competent legal opinion and is of doubtful validity. It seems more prudent to consider such fee schedules as recommendations.

between the client and the engineer. In this situation, the use of fee recommendations, supplemented by the engineer's experience and a careful analysis of conditions affecting the engagement, usually leads to agreement on fees. The test is that they be equitable to both parties and result in a thoroughly professional job of engineering of which both the client and the consulting engineer can be proud.

In business it is always sound to record agreements in written form. Hence, it is common to have engineering service contracts between client and consulting engineer.

chapter **6**

Engineering Service Contracts

Purpose

An engineering service contract serves the same purpose as such a document does in any business transaction. It states the obligations and considerations of the parties involved in the contract. It outlines the services to be performed, the fees to be paid, and the conditions under which the work will be done.

Written contracts help to assure that the contracting parties have had a meeting of minds. Very often two parties talk about a matter and yet fail to fully understand each other. Reducing the agreement to writing helps considerably in obtaining a meeting of minds before work is undertaken. Written contracts minimize later disagreement. When each party can refresh his mind regarding the terms of a signed contract, the chances of later disagreement are reduced. Finally, a written contract provides legal protection to both parties in case of controversy, death, change of personnel, and similar unexpected events. Should controversy arise over services to be performed or fees to be paid, the written contract serves as a basis for legal action. For these reasons, consulting engineers normally use written documents to evidence their agreements with clients.

Form of Contract

There are several forms of written documents that serve as contracts. These include:

1. A formal contract prepared in legal form and executed by the proper contracting authorities.
2. A more simple letter contract, usually prepared by the consulting engineer, a copy of which is executed by the client and returned to the engineer.
3. A purchase order issued and executed by the client, which becomes an agreement upon acceptance by the engineer.

Formal contracts are required by many governmental bodies and, hence, are normally used with such clients. However, a formal contract is desirable on any major engagement. A letter contract is convenient for minor engagements because its preparation is simple and quick. Purchase orders are often preferred by industrial concerns with established purchasing procedures. They are satisfactory if an adequate statement of the conditions of agreement is included in the order or attached to it.

Content

There is great variety in the format, arrangement, and language of engineering service contracts. However, examination of numerous contracts proposed by engineers, clients, and engineering societies reveals little difference in the basic content. These elements are well established both by practice and by legal requirements and may be divided into the following categories:

1. Identification of contracting parties.
2. General recitations.
3. Commitment.
4. Scope of services.

5. Basis of compensation.
6. Provisions protecting the client.
7. Provisions protecting the consulting engineer.
8. Signatures.

Contract provisions are seldom grouped under the categories listed above. However, all categories are needed in a well-prepared engineering service contract whether it is a formal one, a letter contract, or a purchase order. The usual provisions under these divisions are discussed in subsequent sections.

Contracting Parties

The first item in any contract is usually the identification of the contracting parties. The names and addresses of the client and the consulting engineer are stated together with their entities, that is, individual proprietorship, partnership, corporation, municipal corporation, or otherwise.

To avoid repetition of full names throughout the balance of the contract, they are usually identified by a short term. For instance, one may be referred to as "party of the first part" and the other as "party of the second part." Or the consulting engineer may be referred to as "engineer" and the client as "client," "owner," "company," "city," "district" or a similar designation. It is important that contracting parties be correctly identified. This may seem an unnecessary caution but it is surprisingly easy for an error in the clients name to slip through.

Recitations

The identification of contracting parties is usually followed by the recitations of the so-called "whereas" clauses. This term is used because attorneys generally begin these statements with the word "whereas." Such clauses list pertinent facts leading to the contract.

Much of the language contained in "whereas" clauses is window dressing. However, from the consulting engineer's point of view, it is desirable to have a recitation of the client's authority to enter into the contract. This is particularly true if the client is a governmental or quasi-governmental organization. If the consulting engineer has any doubts regarding the client's authority to contract or about the recital of this authority in a proposed contract, he should seek competent legal advice.

Commitment

A statement of engagement of the consulting engineer by the client consists usually of two parts, the commitment of employment and the project description.

The commitment is usually stated in legal language, somewhat as follows:

> NOW THEREFORE, it is hereby agreed, by and between the parties hereto, that the Owner does hereby retain and employ said Engineers to act for and represent it in all engineering matters involved in the project. Such contract of employment to be subject to the terms, conditions and stipulations as hereinafter stated.

The project description serves to define the project or general area of service. When the contract deals with a definite project which will result in construction, the description may refer to the nature of the project, its location, and perhaps its size or capacity. If the project has been the subject of an earlier feasibility study, it may be defined by reference to such study with the report identified by title and date. On the other hand, if the services are not in connection with a construction-type project, there is still need for a definition of the general area of activity. This may be stated as a study or investigation of a specified operating problem in a given manufacturing plant, a valuation of a certain property, or a survey of a given tract of land.

Project descriptions need not be lengthy but should be adequate to provide clear identification.

Scope of Services

After the project is defined, a detailed statement describing the services to be performed by the consulting engineer is needed. If the engagement involves only consultation, studies, investigations and feasibility reports, or valuations and rate studies, the statement may be quite short. On the other hand, if it involves complete engineering services on a major construction project, the statement will be lengthy.

Descriptions of services must be tailored to suit the situation and will be found to vary greatly from engagement to engagement. The following check list indicates some of the specific service items that may be included:

> Consultation.
> Investigation and study.
> Field investigation and observation.
> Preparation of feasibility studies.
> Preparation of economic analysis.
> Preparation of cost estimates.
> Preparation of preliminary designs and plans.
> Detailed design.
> Preparation of plans and specifications.
> Assistance in procurement.
> Advertisement for bids from contractors.
> Assistance in award of contract.
> General supervision of construction.
> Resident supervision and inspection of construction.
> Preparation of assessment plats and schedules.
> Final inspection.
> Testing of equipment.
> Surveys.
> Preparation of valuations.
> Preparation of rate studies.
> Consultation or supervision of operation.
> Assistance in court work.

In order to avoid argument later, it is also helpful to list in the contract certain services that are not included. This is particularly true if the agreed scope of services departs from the normal. For example, if site surveys are to be furnished by the owner rather than the engineer, it is well to exclude these from the scope of work. Other items often excluded on projects involving construction are: property surveys, subsurface borings, laboratory tests of materials, and preparation of special assessments.

Occasionally the scope must deal with optional services upon which decision is to be made later, by the client alone or mutually with the consulting engineer. The scope of such options should be clearly stated, together with the time and method of decision upon them.

Compensation

The contract should set forth the agreed fees and the time and method of payment. Even though a published schedule of fees prepared by a professional engineering society is used, it is desirable that the schedule be fully stated in the contract. In addition, the contract should state the procedure to be followed in making payments. Ceilings or limitations on fee, reimbursement for out-of-pocket costs, and similar items also need to be stated.

With a lump sum fee, little is required beyond the amount. With time charges, whether hourly, daily or monthly, a schedule of rates is required for various classifications of engineers, technicians, and others. If the fee is a percentage of construction cost, the schedule of percentages must be included. A clear definition of construction cost is also required. With direct salary plus a percentage, it is necessary to state the salary rates for different classifications of personnel, together with the percentage to be applied thereto. If such a contract is based on direct payroll cost rather than direct salaries, a definition or formula is needed for determination of direct payroll.

A contract calling for reimbursable costs plus a percentage or a fixed fee requires a thorough definition of the items to be reimbursed. In addition, it must state the basis of the cost plus, either as a percentage or a fixed fee. With a retainer contract, the monthly or annual basis thereof must be stated. It is also desirable to indicate clearly the circumstances under which additional compensation is appropriate and to establish the basis of such additional compensation.

Whenever certain out-of-pocket costs are to be reimbursed, they should be defined clearly. This may be required with any of the several types of contracts. If a ceiling is used with any type of fee, a statement is required of the amount or the method of establishing the ceiling.

On any type of contract, it is desirable to set forth clearly the times of payments and the conditions under which they are due the engineer. Among the various alternatives are these:

1. Payment upon completion of specified services, or portions thereof.
2. Payment upon monthly invoices covering services performed during the preceding month.
3. Payment in accordance with a predetermined schedule expressed perhaps as a specified amount per month.
4. Payment in proportion to progress on the engineering work as certified by the engineer.
5. Payment in proportion with construction progress.

With any of these methods, a percentage—often 10 per cent —of the fee earned may be withheld until final completion.

It is in the consulting engineer's interest to obtain payments progressively with his work and thus to reduce the working capital required to carry a given contract.

Protection to Client

Although the client's fundamental objective is the obtaining of engineering services of the scope outlined in the contract, he is entitled to protection on a number of points. These include the items now presented, all of which are legitimate.

Schedule. The dates of completion of various portions of the work should be stated or a formula for determining rate of progress included.

Insurance. The engineer may be required to carry certain forms of insurance for the protection of the client and may be asked to provide the client with proof of such coverage.

Assignment. The engineer should be prohibited from assigning a contract or subcontracting a portion thereof without the written approval of the client.

Termination. The client is often given the right to terminate the contract when performance by the engineer is unsatisfactory or the project is abandoned.

Personnel. The engineer is required to provide competent personnel and to remove from the work any employee whose work or conduct is unsatisfactory.

Approvals. The engineer is required to obtain certain approvals from governmental agencies or others who have the authority of review and approval.

Records. The engineer is required to maintain accurate records during construction and to provide the client with plans showing final construction.

Reports. The engineer may be required to submit periodic reports to the owner during the progress of the work.

Fee application. Occasionally fees paid for preliminary studies and investigations are applicable against the total fee for complete engineering services.

Protection to Engineer

Although the consulting engineer's fundamental objective is the receipt of fees for services rendered, he is entitled to protection on a number of items. Those listed below are legitimate ones to which the responsible client should have no objection.

Extra work. A basis should be established for authorization and compensation for extra work.

Services not included. A clear-cut statement of services not included is in order, particularly if the scope differs from normal.

Assistance. Engineers should have the right to employ such assistance as is required to perform the engineering work.

Change in scope. The engineer should be protected against any change in scope of the work that may change the magnitude of his work.

Delay. The engineer should be protected against delays beyond his control, or resulting from action of the client.

Records and data. Often the engineer is dependent upon receipt of certain data and information from the client. The client's obligation to furnish these upon a predetermined schedule may be stated.

Client's approval. Where the approval of the client is needed at certain points, it may be requested within a time limit to avoid delay.

Termination. Any clause giving the client right to terminate should include provisions for proper compensation to the engineer for the work completed up to the time of termination.

Signatures

From a sales point of view at least, signatures are the most important part of the contract. They signify the job has been sold, the terms negotiated, and the contract executed. Contracts may be signed by one or more persons for the client, depending upon legal requirements and upon the authorities granted for signatures.

Often, as in the case of corporations, one individual has the authority to sign. Frequently contracts are executed by one individual and attested by another, a practice common with municipal corporations. Occasionally, with a board or commission, signatures of three or more individuals are required.

For the consulting engineer, the method of signature will depend upon the legal entity. With the individual proprietorship, the owner or someone authorized by him will sign. With a partnership, one partner is often authorized to sign for the entire group. With a corporation, one of the officers will usually sign, sometimes with an attest.

Standard Contracts

Some consulting engineers develop standard contracts that they attempt to use. The objection to standard contracts is that conditions vary so much with each job that a great deal of modification becomes necessary. Moreover, many clients, particularly governmental bodies, have standard forms of their own that they insist upon using.

For these reasons I have never found standard contracts worthwhile. Instead, our firm uses "guide" contract forms containing the usual clauses and alternatives. With such a guide form, the preparation of a contract is simplified and the chance for omission of an important clause minimized.

Examples

Neither one nor a dozen engineering service contracts can demonstrate all the circumstances or situations the engineer will encounter. However, in order to demonstrate the form and appearance of engineering service contracts, two samples are presented in Appendix A. One is a formal contract covering complete engineering service on a construction project; the other is a letter contract for a study and feasibility report.

Contract Execution

Contracts, to be binding, must be executed by the authorized representatives of both parties. It is important, therefore, that the engineer be properly advised as to the authorized executors for the client. The consulting engineer should avail himself of competent legal advice in these matters and make certain that contracts are executed by the proper parties.

Moreover, the consulting engineer should ascertain that proper authority has been given to the signers for the client. This is particularly true in connection with municipalities and other governmental bodies where action by ordinance or resolution is required. There are occasions when the execution of the contract has little value unless the governmental body has taken proper steps to budget the expenditure. Unless the engineer watches such situations, he may find that, although he has a legally executed document, no provision has been made for funds.

The consulting engineer is entirely within his rights in requiring certification or proof of authority of the client's representatives to sign a contract. Often a contract must be approved by a third party, such as a governmental agency or financing organization, before it is binding. Unless the engineer is aware of these requirements, he may obtain a properly executed contract only to find some necessary approval missing.

Such precautions may seem unnecessary. Their importance, however, can be proven by a single experience with an irregularity in proceedings that invalidates a contract or prevents collection of fees. Having had a few experiences of this type, I urge careful attention to the legal aspects related to the authorization and execution of contracts by clients. Since the engineer is dealing with legal matters, he should seek the advice and guidance of a competent attorney.

Contract Amendments

Conditions may arise that require a change in a contract with the client. These may be a change in scope of work, in schedule, in fee, or a change in the responsibilities of either engineer or client.

If events appreciably alter the conditions of the contract and, particularly, if they change the compensation, the engineer should insist on an appropriate amendment. Reliance upon verbal understanding or even upon casual exchange of letters is undesirable, particularly if the individual dealt with is not the signator to the contract. Contract amendments can be prepared in several forms, formal documents, letter agreements, or purchase orders. If the changes are extreme, the preparation of a new contract to replace the old may be the simplest procedure.

Whatever the form, it is important that the amendment be properly executed with adequate authority. Cautious attention to these matters will save the engineer some trying moments when he learns that an apparently approved alteration lacks legality.

Substitute for Contract

Sometimes a client refuses to bother with the preparation of a contract, particularly if the magnitude of the services is small. In such cases, the consultant may consider it unwise to press for a contract.

In such circumstances, I have often written a brief letter to the client thanking him for the engagement and confirming the verbal agreement reached. Although such a letter is not a binding contract, it helps to assure understanding and avoid future disagreement. Such a letter can include a request that the client advise promptly if it does not properly reflect the verbal agreement. Such a confirming letter also serves the consulting engineer as a written record of his commitments.

Current practices by which clients select consulting engineers are a compromise between a truly professional approach and strong competitive pressures. Most of the ethical problems confronting consultants and clients are in this area.

chapter 7

Selection of Engineers

Theory versus Practice

Ideally clients should seek out consulting engineers and engage them for their ability and experience. This is the theory of professional practice in the legal and medical professions, one that is often quoted as an example to consulting engineers. Additionally, this is the procedure implied by some of the ethical codes prepared by various engineering societies to govern the conduct of consulting engineers.

In practice, however, the selection of engineers does not follow this ideal, chiefly because of the differences between consulting engineering and such professions as law and medicine. The two latter professions deal largely with personal services performed on an individual basis, usually within a single community. On the other hand, consulting engineering deals with an impersonal service performed principally for organizations involving large projects scattered over an extensive geographical area.

For these reasons most consulting engineers aggressively pursue clients to obtain engineering engagements. Only a comparatively few consultants are so well established and so favorably known as to rely on clients coming to them. Those fortunate few are likely to have small organizations practicing in highly specialized fields and often working principally for other consultants or for architects. In my own experience, less than 5 per cent of the new clients obtained in the past 25 years have come to us without solicitation.

Therefore, to be practical, consulting engineers must concern themselves with the actual competitive situation, seeking always to encourage the highest professional standards. Moreover, engineering societies that desire to raise professional standards need to recognize these competitive facts of life. Only in this way can they effectively influence higher standards of ethics and conduct beneficial not only to the profession but also to clients and the public.

This chapter discusses the processes by which consulting engineers are chosen, the competition they face, and ethical problems related to selling their services.

Competitive Situation

In most cases the selection of consulting engineers, at least on major engagements, is a highly competitive process. The prospective client is usually contacted and solicited by representatives of a number of consulting firms endeavoring to sell services. The competition may be intense even when the practices of all competing consultants are highly ethical. Unfortunately intense competition may lead occasionally to practices that are, at best, borderline and, at worst, flagrant violations of the accepted mores of professional conduct.

Unnecessary competition, and even unethical practice, on the part of consultants is often encouraged by the client. Generally this is unintentional and occurs because the client understands neither the role of the consulting engineer nor the proper procedure for making an intelligent selection of engineers. Unfortunately it sometimes occurs because the client deliberately adopts unethical practices himself, and tries to play one consultant against another.

In addition to the competition between each other, all consulting engineers are in competition with certain suppliers, manufacturers, governmental agencies, and contractors who purport to offer "free" engineering. Such concerns seek to persuade clients to entrust engineering to them rather than to consulting engineers. Consulting engineers also compete with engineering-construction concerns that integrate engineering with construction and offer a turn-key or package job on construction projects. Finally, consultants are, in a sense, in competition with the client's own engineering staff, which frequently resists the idea of outside engineers.

To meet these competitive situations, nearly all consulting engineers carry on sales programs with various degrees of agressiveness and effectiveness. (See Chapter 8.) These activities seek to sell the firm to prospective clients and to assure the engagements a consultant must have to maintain a practice.

Who Selects

The client, of course, selects the consulting engineer, but clients range from individuals, on the one hand, to mammoth corporations or governmental organizations on the other. The selection is made by one or more persons who have been delegated the authority.

If the client is an individual proprietor or a partnership, it is likely that the selection will be made by the owner or by one or more of the partners. In some cases, it may be delegated to a manager or other representatives.

In the case of the corporate client, numerous possibilities exist. Sometimes the decision rests with the board of directors acting upon recommendations of the president or chairman of the board, but more probably the selection will be handled by officers or staff. In that case any one or a combination of several officials may be involved, including the president, vice president, general manager, chief engineer, plant manager, purchasing agent, or others.

With municipalities, the authority normally rests with the city council or similar governing body. However, it may be delegated to a committee, a city manager, a city engineer, or, in the case of a large municipality, to a department head. With state or national governments, the power of selection usually rests with the head of a department or agency. It may be delegated to a selection board or to contracting or engineering officers or staff.

The consulting engineer should correctly determine who makes the selection; otherwise, he may waste his efforts on individuals who have nothing to do with the decision.

Selection without Competition

Selecting engineers without active competition is in keeping with the finest theories of professional practice. In such cases the client bases his decision either upon the reputation of an engineer or upon his previous experience with that engineer.

Occasionally a client who seeks a consulting engineer will select one who has never worked for him and whom he may not know. In such cases the decision is based upon the client's judgment of the competence and reputation of the consultant. His information may come from his own investigations, recommendations of others, or from personal nonprofessional contacts with the consultant. Whatever the source, the client, having made his decision, contacts the consulting engineer and arranges for his engagement.

A more common method of selecting consulting engineers without resorting to competition is found in repeat engagements. In such situations the client, having previously used a consulting engineer and being satisfied with his work, calls him in when another need arises. This procedure is the goal of all consulting engineers—the reward for a job well done. Moreover, it is advantageous to both the client and the consulting engineer.

It is more correct to say such selections are made without apparent or formal competition at the time of selection than to say no competition is present. What actually happens is that the client comes to a decision without interviewing a number of competing engineers. However, any client whose need of engineering services is evident will probably have been contacted by several consulting engineers presenting their qualifications.

Selection with Competition

Active competition is the normal pattern for most important engagements. This may not be the case if the client has a long standing association with a particular firm of consulting engineers. Even so, other competitors are likely to contact the client in the hope that he may change his mind the next time he retains engineers.

The degree of competition depends upon circumstances but, on the whole, it can be very severe. Recently a small midwestern municipality invited sixteen firms to appear before its city council as it selected engineers for a sewage treatment plant. Such practice, which I label an "engineering convention," represents the extreme. The representatives of invited firms were given only 10 minutes to present their qualifications. The presentation was followed by a request for written qualifications and a proposal. Such a practice would be ridiculous if it were not so serious. It seldom results in an intelligent selection of engineers. Aside from its dubious ethics, it is a great waste of time and expense. The fact that consultants participate in such gatherings tends to encourage the method. However, participation is often inadvertent, for an invitation may be accepted without knowledge of the number invited or the procedure planned.

In a more normal competitive selection, the client will probably discuss the proposed engagement with several consulting engineers who call on him. In this process, one engineer may make such a favorable impression that the client will select him when ready to proceed. However, it is more likely that the client, when ready to make a selection, will interview several of the firms that have made contacts or have been recommended. Such interviews will concern the engineer's qualifications, availability, method of procedure, experience, and method of compensation. The client will then request a formal proposal from one or more of the firms. All of this may be done quite informally by private organizations and, in some cases, by municipalities or governmental organizations.

However, many municipalities, counties, and other governmental organizations assume a legal requirement for more

formal competition.* In some instances this is carried to the
extent of asking for competitive bids on the engineering services,
an unethical practice that is objectionable to all professional engineering organizations and reputable consulting engineers.

Normally, however, the practice of municipalities stops short
of competitive bidding. A number of consulting engineers are
requested to appear and to submit proposals covering the required services. After these proposals are received, a decision is
made. Presumably the engineer's experience, competence, availability, personnel, and similar factors are considered, as well
as the fees proposed. It goes without saying, however, that with
such procedures, fees may become the determining factor.
Nevertheless, this procedure, which is widely used to meet actual
or assumed requirements for competition, does allow consideration of factors other than fee.

Selection Board

Another method of selecting consulting engineers calls for the
use of a selection or contract board. The client, perhaps maintaining a file of material and brochures submitted by consulting
engineers, chooses a number of firms he believes to be qualified.
These firms are contacted to determine their interest and a
number, after being informed of the nature and scope of services
desired, are invited to appear before the board.

The board questions the engineers about their experience,
competency, availability, and such matters, but fees are not discussed. After interviews, the board selects a first and perhaps a
second and third choice. It then attempts to negotiate a contract with the first choice. Failing to do so, it moves on to the
second and the third choice if necessary. Many agencies of the
United States government use this method, which has the en-
dorsement of professional engineering societies.

---

* Legal requirements affecting selection of consulting engineers by municipalities, counties, and states are established by state laws. To the best
of the author's knowledge, no such state requirements require "competitive
bids" for the engagement of professional services. Many state and national
engineering societies have obtained legal opinions supporting this conclusion that competitive bidding for selection of consulting engineering services is not legally required.

Suggested Procedures

Whenever possible, consulting engineers should encourage clients to select engineers on a merit basis, using a procedure similar to the following:

1. The client should select for interviews one or more firms qualified for the particular engagement under consideration.
2. The client should interview separately those selected, discuss the nature and extent of services required, and question them regarding experience, qualifications, personnel, and related matters.
3. The client should then consider the relative qualifications of the engineers interviewed and make such further investigations as are desired.
4. The client should then select the firm it desires and enter into negotiations regarding fees and contractual relations. Should no agreement be reached with the first choice, the client should terminate negotiations and consider a second choice.

This procedure, besides being fair to consulting engineers, serves well the interests of clients and public.

Factors to Be Considered

In comparing the relative merits of consulting engineers, a client may well consider the following factors:

1. *Experience.* What is the engineer's experience, both in general practice and in the specialized field under consideration?
2. *Concept of problem.* Does the consulting engineer display a good understanding of the client's needs and problems?
3. *Competence.* Does the previous work of the consulting engineer demonstrate adequate competence to handle the work under question?
4. *Staff.* Does the consulting engineer have adequate staff or is he planning to engage them after he gets the work? How does his staff compare with the work load he has?
5. *Knowledge of local conditions.* Is the consulting engineer

familiar with the local conditions and situation under which the client's work will be done?

6. *Integrity.* Has the engineer a reputation for outstanding integrity and honesty?

7. *Cooperation.* Is the engineer wholeheartedly cooperative with the client?

8. *Standing.* What is the engineer's standing in his profession?

9. *Reputation.* What is the consulting engineer's reputation with existing clients?

Competition from Other Sources

Consulting engineers are confronted with a great deal of competition from outside the profession; some of this competition is legitimate and some is not.

Many manufacturers and material suppliers offer allegedly free engineering service to concerns having engineering problems. Naturally the manufacturer or material supplier should design the product and provide engineering data to those who use it. Often, however, he goes beyond this role and provides engineering services that should come either from the owner or his consulting engineer. The staff of a manufacturing concern owes its basic loyalty to that organization. Its members are not in a position to give unbiased and objective advice. A client who depends upon such free engineering does so with a twofold risk. First, the "free" engineering undoubtedly will be slanted to favor the use of that manufacturer's produce, and, second, an entirely inadequate job of engineering may be provided.

The same situation prevails with contractors who offer to furnish an integrated engineering and construction job. Such an approach, known as the "turn-key" job, is frequently encountered in the industrial field. It is offered by many companies, including some of the large engineering-contracting concerns who sell their projects, not so much on the basis of "free" engineering as on other alleged advantages, such as cost, coordination, and speed, of integrated engineering and construction.

Although there are situations in which these advantages seem to warrant a turn-key job, the spread of this system deprives the consulting engineer of a market for his services and the client of an impartial check on the contractor.

Education of clients is the only way to meet these types of competition. The consulting engineer must convince the client of the advantages of using a consulting service. The more adequate engineering services of the consulting engineer will normally result in:

1. A lower over-all cost of the project due to checking and control of the contractor and less "over-design."

2. Savings in cost, resulting from better competition.

3. A more serviceable finished project.

4. More objective and impartial consideration of the client's needs and representation of his interest.

Competition from Clients

In a sense consulting engineers are in competition with the staff of any client who has an engineering organization. Certainly every concern has the legitimate right to maintain a staff to meet its own engineering needs. Yet such an engineering staff often resists the engagement of outside consulting engineers. It may do so fully believing itself capable of providing a better and more economical service for its organization, or because it has a prejudice against consulting engineers, or because it seeks to build a bigger department or empire.

If the consulting engineer is to obtain engagements with such organizations, he must sell the advantages of using his service. Some of his selling points include his broader experience, his capacity to put an adequate staff to work immediately, the shouldering of responsibility, and the specialized knowledge he brings to the problem. All of these are more fully discussed in Chapter 2 under the heading "Why Clients Employ Engineers."

One of the principal contentions he must meet is the question of cost. It is usually possible to show that the services of a consulting engineer will save the client money. This can be done provided the client compares a service equal in quality and extent to that proposed by the consultant, and provided the client's analysis reflects all items of cost. Often the client's engineering staff contemplates a lesser amount of service than that planned by the consulting engineer. Often, too, the client's accounting system fails to charge full overhead costs against its engineering department.

Role of Competition

Competition is here to stay in the consulting engineering profession. There is no remote likelihood that engineers will lessen their activities to solicit work in the foreseeable future.

Personally I favor competition; it is the lifeblood of a free enterprise system. It compels established firms to keep alert, seeking always to give better service. It permits the new engineer to break into the field. I believe it to be in the best interests of the client, the public, and the profession. However, consulting engineering is a profession—not just a business. Hence, it is important for competition to be on a highly ethical level. This is the challenge that faces consulting engineers and their professional organizations. To meet this challenge, the consulting engineering profession must do three things. First, it must raise ethical standards among its own members. Next, it must educate clients and the public regarding the role of the consulting engineer and the ethical aspects of his practice. Finally, it must police itself against unethical practice.

Ethical Principles

Many codes of recommended practice have been prepared by professional engineering societies. The more common precepts relating to competition among consulting engineers are stated below.

The consulting engineer in competing with others of his profession shall:

1. Uphold the reputation of the profession and protect it from misrepresentation and misunderstanding.

2. Refrain from seeking to displace a competitor after becoming aware that definite steps have been taken toward his employment by a client.

3. Refrain from price competition in negotiating fees and contracts for engineering services and avoid such practices as: (a) Reducing his usual fee after being informed of the charges named by another. (b) Attempting to secure professional work awarded on the basis of competitive bidding. (c) Offering free services with the hope of securing an engagement.

4. Solicit engagements for engineering work according to high professional standards, refraining from using improper or questionable means, and declining to pay a commission for engineering engagements.

5. Avoid advertising in a self-laudatory manner or misrepresenting his qualifications, competency, staff, or availability.

6. Avoid injuring directly or indirectly the professional reputation, prospects, or practice of another engineer. (However, if he considers a competitor guilty of unethical, illegal, or unfair practice, he will present the information to the proper authority for investigation and action.)

7. Refuse to review the work of another engineer except with the knowledge and consent of the other engineer unless the employment of such engineer has been terminated.

8. Avoid association with consulting engineers who do not conform to proper ethical practice.

9. Refrain from using the advantage of a salaried position to compete unfairly with other consulting engineers.

Unethical Practice

Many cases of unethical practice result from misunderstanding or inadvertence. Sometimes the heat of intense competition to obtain a particular engagement prompts an engineer to action he normally would neither approve nor follow. Sometimes, also, a client unknowingly encourages steps on the part of consulting engineers that are breaches of good ethical practice. Both consulting engineers and clients should strive to keep the procedures of selecting engineers on a high plane of ethical practice that will benefit both.

Most of the cases of unethical practice in connection with selection of engineers or competition between them fall into five categories:

1. *Fee competition.* Violations are less in the area of "competitive bidding" than in reducing usual fees to obtain a competitive advantage after a proposal has been submitted. Competitive bidding has been successfully opposed by various engineering societies, and progress has been made toward reducing this evil. When such a situation is encountered, a reputable consulting engineer will not only withdraw his name from consideration, making his objections known to the client, but will also request an appropriate engineering society to protest to the client.

The most common violation of price competition is lowering a fee after the engineer has submitted a proposal. This action may be requested by a client but sometimes is invited by an engineer anxious to obtain the engagement. Once a fee is proposed, it is unethical to lower it unless the client reduces the scope of the service.

Another form of price competition is the setting of unreasonably low fees that do not permit a competent job of engineering. A consulting engineer should be guided by the various recommended fee schedules unless his analysis and experience indicate a different fee is in order. It is certainly unethical to propose a fee that will not permit a competent and complete job of engineering, and it is unwise for a client to seek or accept such a fee proposal.

2. *Misrepresentation.* A number of unethical representations may be made to a client. One is to fail to advise the client of any associations or connections (such as with a contractor or manufacturer) that in any way prevent the consulting engineer from objectively and honestly representing the client's interests. Another misrepresentation is overstating the consulting engineer's experience, qualifications, and staff. Occasionally we encounter the unethical practice of misrepresenting probable construction cost of a project. By suggesting an unusually low estimate of cost, a competitor may unethically influence a client.

3. *Advertising.* For decades the only acceptable and ethical advertising by consulting engineers has been the professional card carried in the engineering or technical journals. In recent years, however, some consulting engineers have carried display advertising in business magazines. A review of a recent issue of *Fortune* magazine reveals a number of display advertisements for engineering services. Some of the ads are full page and a few are in color. Although the majority advertise companies offering construction services as well as engineering services, all of them purport to offer consulting engineering. This practice certainly is "self-laudatory" and violates the long-established precepts of ethical practice. Therefore, reputable consulting engineers refrain from such advertising.

4. *Sales method.* Certain sales methods are unethical. One of these is to criticize or undermine a competitor's reputation or ability. In addition to being unethical, it is a poor sales method, for the best approach is always to sell one's own wares. Excessive entertainment is in bad taste and unbecoming to a professional approach. Unfortunately a few engineers seem to depend largely on entertainment as a sales device.

5. *Commissions.* Payment of commissions for engineering engagements is an unethical practice. Obviously a consulting engineer has need for sales contacts and may, under certain situations, legitimately engage those who are not engineers to represent him. But the use of the "five percenter," the "finder" and other influence peddlers to obtain engineering engagements is certainly unethical on the American scene.

Raising Standards

Both the consulting engineer and the client have an interest and a responsibility in obtaining higher ethical standards to govern the selection of consulting engineers. Such higher ethical standards will assure better engineering service. This need prevails in spite of a substantial rise in ethical standards in the last two decades. It is so even though the great majority of consulting engineers practice ethically and discourage actions unbecoming to the profession.

This responsibility for achieving and maintaining high ethical standards is a joint one; it rests on the consulting engineer as well as the client.

To build a clientele, a consulting engineer must sell his services to prospective users. Without engineering jobs, he will soon be out of business.

chapter **8**

Selling Services

Sales Activities

To obtain engagements in the face of the competition described in Chapter 7, most consulting engineers maintain a continuous sales effort to acquaint prospective clients with their talents and availability. Various commentators on the consulting engineering profession object to such terms as "sales program" and "selling." Some cling to the illusion that clients come to consulting engineers without any activity on the latter's part. Others, recognizing the competitive situation, suggest such terms as "contact" and "promotion" to label the sales activity. However, it is preferable to use the term "sales" to describe these activities. Such frankness is a necessary first step in placing this activity on a higher ethical level.

Sales activities consist principally of contacts with prospective clients, supplemented by legitimate advertising and ethical use of sales aids and publicity. Such a program is augmented by the recognition and reputation of the firm, based upon its competence and experience. Add the element of luck and you have the ingredients that a consulting engineer uses to build a clientele.

Reputation

No single factor plays a greater part in the success of a consulting practice than a good reputation. Unless a firm is reputed to be competent and skilled in its work and to possess a high level of integrity, it cannot long survive, regardless of its sales activities. The soundest way to build a good reputation is to give the kind of service that results in satisfied clients. Thus, any consulting engineer seeking a successful practice must, above all else, execute well the services for which he is engaged.

There are other means by which a consulting engineer and his staff may help build a reputation. One of the most effective is to write and publish professional and technical papers. These bring the consultant's name before other engineers and clients in a highly professional manner. They also offer an opportunity to establish a reputation for technical competence in selected fields. Participation in engineering societies at the local, state, and national levels is helpful. It brings the consulting engineer in contact with his competitors, other engineers, and, frequently, with clients. The contribution he makes, the leadership he exerts, and the conduct he displays all contribute to advancement of the profession as well as to his reputation. Finally, a consulting engineer builds a reputation in his nonprofessional activities. Participation in service clubs, churches, and community activities brings him in contact with people. Even though local activities may touch very few clients, they help establish his reputation. His financial affairs also affect his reputation.

Reputation—professional and personal—has a way of spreading among competitors, salesmen, engineers, and clients. Moreover, the prospective client is likely to investigate an engineer's reputation in his home town, among his clients, and elsewhere.

Repeat Engagements

Every consulting engineer wants the maximum number of repeat engagements from the clients he has served. Repeat engagements are a potent factor in building a reputation, for they signify that former work was completed to the satisfaction of the client. The percentage of repeat engagements enjoyed by a firm is one clear measure of the quality of its engineering work and its human relations. If the percentage is not satisfactory, the consulting engineer should scrutinize his work carefully because he is failing in some respect.

Although superior performance is the basis for repeat engagements, the consulting engineer also needs to maintain contact with former clients. Often there is a long period between a client's needs for engineering services, during which the consultant should not rest his case entirely on previous performance. By keeping in touch with the client he learns of coming needs and reminds the client, from time to time, of his availability and interest.

Advertising

In the sales field, generally, there are two basic methods of carrying a message to a prospective purchaser, advertising and personal contact. However, the accepted ethical codes for consulting engineers severely limit the use of advertising.

It is accepted practice that consulting engineers shall not advertise in a self-laudatory manner. This prohibits display advertising and limits consulting engineers to advertising in publications through "professional cards." A professional card is a listing in a directory of consulting firms in an engineering journal or trade magazine. Professional cards, small in size, carry the name and address of the firm, often with a listing of its principal fields of endeavor.

A survey by *Consulting Engineer,* published in February, 1959, indicated that 47 per cent of consulting engineers carry one or more professional cards. Some of these are merely courtesy cards supporting a local engineering society publication, but many are placed in national publications. Occasionally cards are carried in newspapers. The principal value of professional cards is prestige or institutional advertising, which keeps the firm's name before other engineers and prospective clients. Few engagements are directly traceable to a professional card. However, now and then, one receives a letter or a telephone call from a prospective client traceable to such a card. The cost of professional cards varies with the frequency of publication and the circulation of the magazine. The extent of the use of professional cards depends upon the consulting engineer's judgment of their effectiveness in reaching the clientele he seeks, directly or indirectly.

Closely related to professional cards are listings in the classified sections of telephone directories. These are of value in the community in which the consultant is located.

Contacts

Since ethical advertising permitted the consultant is not an effective device for selling services, the consulting engineer must rely primarily upon contacts with prospective clients. Such contacts, supplemented by recommendations from satisfied clients and supported by a sound reputation, are the source of most engagements. This is essentially the same process by which other professional people obtain clients. It is illustrated by the young lawyer, endeavoring to build a practice, who joins many organizations and participates in many community activities in order to become known to more people. The consulting engineer has the same need of contacts, but is faced with a much more difficult situation.

He has a geographical problem, for most consulting engineers must draw their clientele from a much wider area than the home town. Next, his clients are not individuals who may be met casually at community functions, but are predominantly corporations and governmental bodies. Finally, many prospective users of consulting engineering services are not aware of the availability of such services and the advantages of using them. Any person with a serious illness will think of the doctor, but the organization with a difficult engineering problem may not think of the consulting engineer. Of necessity, therefore, consulting engineers substitute for the informal contact of the lawyer or doctor a more formal and extensive program to reach clients.

Consulting engineers prepare lists of prospective clients and visit them at their place of business. These contacts serve to locate prospective work and to sell clients on the consulting engineer. Another purpose is to acquaint future clients with the function and role of consulting engineers. Thus, the consulting engineer seeks to widen the market for his services. All of this constitutes solicitation of business and should be recognized as such.

The initial contact with a prospective client is for the purpose of getting acquainted. The consulting engineer trys to meet one or more of the officials or staff who have the responsibility for engineering work. He attempts to learn of the client's present or future needs for consulting engineering services. He makes certain he has the proper contact with the client. Finally, he acquaints the prospective client with his firm and its qualifications, perhaps leaving a brochure or other literature.

Successive calls have the dual function of keeping the consulting engineer's name before the client and of learning more about the client's needs. Usually a number of follow-ups are necessary before the consulting engineer can assume, with any degree of certainty, that the client will think of him if he has a need for consulting services. Such is the fundamental objective of contacts, for unless the client calls in the consulting engineer, work may be placed elsewhere before the engineer is aware of a pending decision.

Once the consultant learns of an immediate requirement, either through a call from the client or by a personal contact, his approach changes from the general to the specific. Concerned now with a definite engineering requirement, he supplements his general qualifications with specific ones applicable to the immediate need. He may prepare a special presentation or provide other literature and data. His activities are now aimed at convincing the client that he is capable of handling the specific project.

Such a program of intensive personal contacts is both difficult and costly. It calls for good organization, sound planning, and an imaginative approach. It should be carried out by well-qualified members of the consulting engineer's organization.

I believe sales and contact work should be done by engineers and never delegated to salesmen, who are not professionally qualified. Furthermore, I think the sales responsibility should be assigned to the upper levels of staff, such as principals and key employees. Such steps aid in maintaining the professional approach to sales. The organizational problem of assignments for contact and sales work is further discussed in Chapter 12.

A well-executed contact program can be effective, a poor one can be discouraging and virtually a wasted effort.

Telephone and Mail

In addition to the personal visit, two other useful types of effective communication with clients are telephone and mail.

After an acquaintanceship has been established with a prospective client, the telephone is a handy and economical device to maintain contact. A telephone call can be made when personal contact is neither warranted nor possible. Used intelligently, it can supplement the personal contact.

Mail contacts can also be used, especially after the initial contact. Personal letters are useful in maintaining communication, arranging appointments, and forwarding or getting information from a client. General mailings to lists of prospective clients may be helpful. These can be used to mail reprints of articles, photographs or descriptions of projects, and other information. However, such direct mail "advertising" is not very effective unless a personal acquaintance has been made. Direct mail material needs to conform to the ethical requirement that it should not be self-laudatory.

Group Appearances

Often the consulting engineer has the opportunity to present his qualifications to a group representing a client: a board of directors, a city council, a contract board, or several members of the client's staff. The group may include three or four, or a dozen or more people. Such opportunities are welcome because the consulting engineer can reach several people at once. However, such an opportunity imposes the obligation to know what he is going to say and to say it well. He should prepare in advance his remarks and the materials he wishes to present: charts, slides, film strips, typical plans, specifications or reports, photographs, brochures, or other material.

He should ascertain the areas of group interest and confine himself to them. Furthermore, he should respect any time limit given him, and allow some of the scheduled time for questions from his listeners. Such group presentations can be of great benefit to a consultant. However, if he muffs the opportunity and fails to win the respect of the group, he may talk himself right out of an engagement. I know this can happen for I have done it myself.

Brochure

Most consulting engineers prepare a brochure for prospective clients to acquaint them with the firm's experience and qualifications. Generally brochures deal with experience, personnel, and qualifications. Such brochures vary from the small and simple to the large and elaborate.

Experience and qualifications are usually presented by photographs of projects that the firm has handled as well as by lists of clients and engagements, grouped by types of work. Sometimes the descriptive material includes costs of projects, operating characteristics, and design features. Personnel may be presented by photographs, lists of employees, and biographical sketches. Normally photographs and sketches are limited to the owners, principals, and key employees. The work and expense of preparing a good brochure is substantial. The content must be determined, the text edited, the layout prepared, and the photographs obtained. Advice and assistance from persons skilled in layout and format are often desirable.

Properly prepared brochures are effective in presenting an engineer's qualifications and in educating clients on the role of the profession. Probably three out of four consulting engineers use some type of brochure, ranging from the expertly prepared printed booklet to the typewritten statement of a few pages. The preparation of every brochure raises again the problem of ethical advertising. Brochures should be factual presentations of experience, staff, and facilities, without self-laudatory comments or claims.

Sales Aids

To borrow a term from the marketing field, there are sales aids that can help a consulting engineer. In place of a brochure, or in addition to one, the following items may be useful in presenting an engineer's qualifications, experience, and work:

1. Lists of engagements showing the date, client's name, and a brief description.
2. Project descriptions outlining the important features.
3. Photographs of structures, buildings, projects, or machines.
4. Lists of personnel, including biographical sketches of at least key personnel.
5. Typical reports, plans, and specifications.
6. Strip films or slides made from photographs of projects.
7. Reprints of professional articles by members of the consulting engineer's staff.

Special Presentations

A specially prepared presentation can be effective if tailored to fit the needs and requirements of a specific client. It is a statement of the engineer's qualifications to handle a specific engagement. Such a presentation may well include data taken from a general brochure, together with other photographs, lists of projects, lists of personnel and similar material. A special presentation may outline in greater detail the engineering features of projects previously handled by the consulting engineer that are similar to the one contemplated by the client.

Publicity

Newspapers, professional and technical magazines, radio stations, and other news media are, of course, always looking for news. Some activities of the consulting engineer are newsworthy. Major engagements, particularly by governmental organizations, and completions of construction projects are news. Unique and imaginative designs of building structures, projects, and machines are news. Promotions, changes in organization, and personnel additions are news.

Such items present opportunities for the consulting engineer to achieve recognition for his work and to keep his name before the public and prospective clients. But, if a consulting engineer wishes to avail himself of such publicity, he must take the initiative in getting information to the news media. He should acquaint himself with the kinds of news stories they want and the mechanics of providing them. Then, by furnishing newsworthy material, he can often achieve legitimate and effective publicity.

Some consulting engineers regularly attend to such publicity and thus obtain broader recognition. A few employ public relation firms or experts to handle publicity. In either case, careful attention should be given to ethical principles. Publicity—if loud, vulgar, or extravagant—can violate the prohibition against self-laudatory advertising. But dignified and factual publicity is ethical and helpful both to the particular engineer and to the consulting profession.

References

Frequently a prospective client will ask persons familiar with the work of a consulting engineer for references and occasionally will check such references. It is always appropriate for a consulting engineer to submit names of individuals who are familiar with his work. These should be responsible officials of clients whom he has served. Moreover, there are times when a consulting engineer is well advised to encourage a client to use these references, particularly if he is not well known in the area where he is seeking work or if his experience and competence have been questioned.

Visits

The consulting engineer, having pride in his completed work, will hope that the client is sufficiently interested to visit some of his projects. Such visits allow the client to see the work of the engineer and learn of his qualifications from former clients. A visit to the consulting engineer's office can also be revealing. Would that such visits were always made by clients! They would serve to reduce exaggerated statements of a firm's staff and organization.

Proposals

At the appropriate time the consulting engineer will submit a proposal to the client. If the selection has been made by the recommended method (as discussed in Chapter 7), the engineer will already have presented his qualifications and the client will be ready to negotiate fees.

Proposals can be presented in several ways. The simplest is a letter stating the schedule of fees and the more pertinent features of the proposed contract, or, the fees may be presented verbally, to be confirmed by contract documents after the negotiation is complete. Another approach is to prepare a draft of the engineering service contract and to use it as a proposal. Such a draft not only states the proposed fees, but fully outlines all contract terms. The choice of these methods will depend upon the circumstances. Often a draft of contract is favored because it clearly presents all conditions relating to the negotiation. Sometimes it may be desirable to accompany the proposal with the firm's brochure or other data outlining qualifications.

It is always desirable to present a proposal personally to the client and, if possible, to go over it with him. This not only avoids misunderstanding, but gives the consulting engineer an opportunity to press the negotiation to a decision. Most of us have had the experience of mailing a proposal in the belief a client had decided to engage us, only to find some competitor had visited him in person and obtained the engagement.

Selection of Clientele

The consulting engineer who is beginning his practice has little choice of clientele. He will take any legitimate client he can get. But, as his practice develops, he should carefully consider the type of clientele he solicits. A good clientele is just as important to the consulting engineer as a good engineer is to the client. The consulting engineer should seek his clients in the fields of engineering in which he is most interested. Moreover, he should seek clients who are cooperative, willing to pay fair fees, and who respect the role of the consulting engineer.

All of us who have practiced for a long time have a list of clients for whom we prefer not to work. These are clients who have been uncooperative, unreasonable, or unwilling to pay promptly. Instead, we seek repeat engagements from the clients with whom we have enjoyed working.

In every field and in every locality, there are certain key organizations with considerable influence among other prospective clients. Naturally a consulting engineer seeks to add them to his clientele.

Salesmanship

Competition among consulting engineers centers on two elements, qualifications and salesmanship. As a client narrows the field of engineers who have contacted him, he will eliminate those he considers least qualified and least impressive. Unless the client makes an extensive investigation at this stage, his decision will be based primarily upon his impression of the various firms as presented by their representatives. This is where salesmanship comes in. A good presentation done in a skillful manner will leave a favorable impression upon a client, even though that particular firm's qualifications are no better than others.

Salesmanship and presentation cannot create qualifications for a consulting engineer, but the best qualified firm can get lost in the shuffle if its case is not skillfully presented.

*The professional relationships of
the consulting engineer extend beyond his clients to the public and
to all those with whom he comes in
contact. Besides his individual contacts, he works through his professional organizations to nurture
higher professional attitudes.*

chapter 9

Relationships

Importance

Human relationships, important in any business or activity,
are doubly important to the consulting engineer who has nothing
to sell except his time and talent. He is judged not only on the
quality of his engineering work and his technical decisions, but
also on his conduct with the many people with whom he deals.

He has a close personal relationship with the officials and
staff of his client. He frequently represents his client in dealings
with governmental agencies having regulatory or approval powers
over engineering matters.

Most consulting engineers work with contractors, manufacturers, and material suppliers who perform services or furnish
items to the client. He also contacts them to obtain information
and data on equipment and materials. The consulting engineer's
relationships with other engineers include those with his competitors as well as those with his employees. Finally, as a professional man, the engineer has a unique obligation to the public.

In all of these areas, the successful consulting engineer needs to develop truly professional attitudes and actions.

What Is a Profession?

The term "professional" is much abused and misunderstood. It has always been assumed that the practice of a learned profession carries a responsibility to society that transcends personal gain. Formerly only theology, education, law, and medicine were considered learned professions. But as society has become industrialized and more highly organized, numerous trades and occupations have endeavored to achieve acceptance as professions. As a result, the term now has a meaning extending beyond those four professions.

Engineers have long contended that their occupation is a profession, and engineering societies have worked to achieve such recognition. Registration of engineers has done much to dignify the use of the term "Professional Engineer" and has given it a legal status. The achievement of a truly professional status for all engineers, however, is a goal yet to be accomplished. The overwhelming majority of engineers are employees of corporations or of governmental organizations. Many combine such functions as administration, sales, and management with engineering. These conditions make it difficult to attain true professional stature and public recognition.

But consulting engineers are different, for as private practitioners they function more in the manner of attorneys and doctors. Whether or not engineers as a whole ever attain professional stature, those in private practice now have it to a substantial degree. Professional attitudes and actions, therefore, should govern all consulting engineers.

Professional Attitude

The statement of professional precepts is a difficult task that has challenged the committees of numerous engineering societies. Many codes and canons of ethics have been prepared to govern the engineer's professional conduct.

A professional engineer has responsibilities extending beyond self-gain and personal recognition. One of the best statements of these obligations is the "Engineer's Pledge of Service" adopted by the Iowa Engineering Society in 1942 as part of its Code of Ethics. The four elements of this pledge are stated below, slightly rephrased to apply to the consulting engineer:

1. The consulting engineer will place service to mankind above personal gain and use his engineering knowledge and skill to benefit humanity.

2. The consulting engineer will render faithful, professional service to his client and honestly represent his interests.

3. The consulting engineer will be governed by the highest standards of integrity, fair dealing, and courtesy in his relations with others.

4. The consulting engineer will encourage the development of the engineering profession and contribute to the improvement of the service of engineering.

The ethical relationships between the consulting engineer and his client have been outlined in Chapter 4, and those affecting competition with other engineers in Chapter 7. Ethical principles governing relationships with others are outlined below.

Contractors and Suppliers

Most consulting engineers work with contractors, manufacturers, and material suppliers. These contacts may include the award of contracts, the purchase of material or equipment, the supervision of construction or installation, and the testing and acceptance of machinery and plants.

The consulting engineer has a quasi-judicial relationship that imposes on him the responsibility to deal fairly with both the contractor and the client. As arbiter between the client and the contractor, he must act with the highest degree of fairness and honesty. The accepted ethical responsibilities of the consulting engineer toward contractors, manufacturers, and material suppliers are:

1. The consulting engineer will conduct purchase negotiations and contract lettings in a fair manner.

2. The consulting engineer will prepare plans and specifications so they are definite and specific.

3. The consulting engineer will not require a contractor to furnish materials or do work not clearly called for in the contract documents.

4. The consulting engineer will require full and complete performance of the provisions of the contract, plans, and specifications.

5. The consulting engineer will promptly furnish the contractor with information reasonably necessary to carry out the project.

6. The consulting engineer will promptly and fairly interpret the plans and specifications when required.

Beyond official relationships with contractors, manufacturers, and material suppliers serving a client, the consulting engineer has other contacts with them. Frequently the consultant calls on such organizations to obtain information he requires in his studies, reports, and designs. To maintain friendly relationships, he should be careful to deal fairly and confidentially with information supplied, and to avoid unnecessary requests.

The consulting engineer has a selfish interest in maintaining good relations and a fine reputation with representatives of contractors, manufacturers, and suppliers. Such persons sometimes provide leads to future work. Moreover, because they travel about contacting many clients they have a real influence on the reputation of an engineer. The consultant's relationships with such concerns, however, must be on a high ethical plane. They should not be encouraged to feel that information given the consultant entitles them to an advantage.

Other Engineers

A consulting engineer has contacts with many different groups of engineers. His relationships with his competitors have been discussed in Chapter 7. His relationships with employees are dealt with in Chapter 13. These relationships with the engineering profession are both personal and composite.

He should endeavor to protect the engineering profession collectively, to strengthen it, and to enhance its stature in the eyes of the public. Every consulting engineer should participate in the work of engineering societies, both professional and technical. He owes it to the profession to contribute to the technical press by preparation of papers and articles and by exchange of engineering information. He can encourage and help students and young engineers to attain an engineering education. He should recognize properly qualified engineers as members of the profession and extend to them courtesies and considerations consistent with his practice. He can help all engineers by informing the public about the profession and by helping to raise the standards of engineering performance and ethics.

In these ways he contributes to the engineering profession.

Relationship with the Public

The engineer has relationships with society, with governmental agencies and bodies having power of authority and regulation affecting his engineering work, and with numerous individuals as such. As a professional man, his responsibilities to society require him to place service to mankind above personal gain and to use his engineering knowledge and skill to benefit humanity.

The engineering profession as a whole has a great opportunity to promote public welfare by safeguarding life, health, and property and by improving the social and economic status of mankind. A reputable engineer avoids association with enterprises or projects contrary to public welfare or that are of questionable, speculative, or illegal character. Both in his professional capacity and as a private citizen, he can often help the public to arrive at a correct understanding of the technical phases of public questions. In his relations with governmental agencies and individuals, the consulting engineer should be governed by the highest principles of integrity, courtesy, and cooperation, qualities that will reflect upon him not only as an individual but also as a member of an honored profession.

Professional Recognition

Professional recognition and stature is something to be earned, not conferred. The consulting engineer has the opportunity to achieve such recognition. Recognition will be earned not alone by technical competence and achievement, but, more importantly, by personal conduct and by acceptance of the ethical responsibilities of a professional man. No amount of legislation, promotion, or education can confer professional status upon an individual. He can win it only by his own efforts. This is the challenge and the opportunity facing all consulting engineers.

Group Action

There are two areas of development that are beyond the capacity of the individual consulting engineer and require group action. One is the continuing need for education of clients, which aims at expanding the market for consulting services. The other is the need for improving ethical and professional standards within the consulting engineering profession. To meet these needs, consulting engineers turn to their professional organizations.

Client Education

A crying need exists for more effective educational programs directed at the client in particular and the public in general. One objective of such programs is the broadening of the market for consulting engineers through greater understanding of their role, function, and value. A second objective is the improvement of relationships between clients and consulting engineers in regard to selection, fees, and working arrangements. The responsibility for executing such educational programs rests both with consulting engineers individually and with their professional organizations.

To date, nearly all that has been accomplished in this area has been done by consulting engineers through personal contacts with present and prospective clients, and there is no reason to expect this situation to change. Although effective publicity and educational programs are beneficial, the main responsibility continues to rest on individual consulting engineers. Clients, who have little interest except when selecting an engineer, are too scattered and diverse to be reached effectively by mass media.

However, professional organizations do have an important part to play in this program. They can prepare and distribute basic information presenting the role of the consulting engineer, his relationship to the client, and data regarding fees, selection of engineers, and principles of professional ethics. Moreover, they can be helpful in the publicity field, and in work with organizations and associations whose membership comprises prospective clients, that is, municipal leagues and trade associations of industries.

Professional Standards

There is a continuing need for higher professional standards, a need that includes both the ethical relationship between competing consulting engineers and the nurturing of higher professional responsibility to clients. The responsibility for obtaining higher standards rests squarely on the consulting engineers themselves and their professional organizations. No one else can or will assume the responsibility. Two approaches are needed to achieve such higher levels, education and disciplinary action. Both of these require group action to supplement individual efforts.

Educational efforts within the profession should be directed toward a broad dissemination of information on professional ethics and responsibilities. These subjects are excellent ones for discussions, articles, and papers that can be encouraged by engineering societies. In the long run, the educational approach will be successful because the unprofessional conduct of consulting engineers is usually the result of ignorance or misunderstanding. Nevertheless, situations do arise that require disciplinary action lest the whole profession suffer from the misdeeds of a few. Hence, many engineering societies have ethics committees that concern themselves with such problems.

The various state boards of engineering examiners are sympathetic to the improvement of ethical standards. They may revoke a license to practice professional engineering when incompetence, fraud, or dishonesty are proven. In some states, they can void a registration for unprofessional or unethical practice. However, state boards seldom take the initiative. In fact, in most states they cannot act until charges are submitted to them. Before disciplinary action can be started, charges must be submitted by other consulting engineers or by their professional organizations. Where such action is required, it can best be initiated by an organization representing the consulting engineers. This method is desirable both because individuals hesitate to file charges and because the group action carries more weight. For these reasons, consulting engineers should encourage their engineering organizations to assume a responsibility for policing the profession.

Many cases of unethical practice, however, can be eliminated by understanding and education. Committees on ethics can be effective in clearing up such cases by consultation with offenders.

Professional Organizations

Consulting engineers have long felt a need for organizations that would accomplish these group activities in the fields of education and discipline. Many individual consultants have given extensive service and leadership to state and national engineering societies in an effort to raise professional levels.

For many decades consulting engineers assumed that the older engineering societies (American Society of Civil Engineers, American Institute of Electrical Engineers, American Society of Mechanical Engineers, and others) or the various state societies would perform these policing and educational tasks. The results have been disappointing because consulting engineers are a minority group within such societies. Moreover, even professional engineers are in a minority in some such organizations. Hence, it is understandable that these groups have not consistently and effectively provided the leadership in the consulting field.

Therefore, consulting engineers have developed their own organizations in order to give greater attention to their particular problems. The American Institute of Consulting Engineers has long been concerned with professional problems of the practicing engineer. However, the membership of this organization is so small (representing only about a hundred firms in 1958) that its influence has been limited.

Within the last 25 years, the National Society of Professional Engineers has gained in size and strength. Concerned only with registered professional engineers, it has had a higher percentage of consultants among its membership. This has permitted greater attention to the problems of the consulting engineer by NSPE and by the state organizations affiliated with it.

Recently the Consulting Engineers Council was organized to meet the widespread desire for an active organization interested only in the consulting engineer. It now has many state organizations associated with it. As its membership is restricted to owners and employees of consulting practices, it truly represents the consulting engineer. State organizations of the Consulting Engineers Council are frequently competing with state branches of NSPE, some of which have private practice sections representing the consulting engineer. This competition can be healthy and will result in greater activity. From it should emerge more effective action regarding the interests of consulting engineers.

All consulting engineers have a definite responsibility to support actively the organizations that are working to develop a more professional attitude on the part of the public, clients, and consulting engineer. Such concerted action, supplemented by individual effort, will enhance the professional status of consulting engineers.

Management of a Consulting Engineering Practice

part two

The first part of this book was concerned with the role of the consulting engineer and his external relations with clients and others. With this as a background, Part Two deals with the internal problems of a consulting practice. It discusses the many areas of organization, personnel, plant, facilities, procedures, and management with which the consulting engineer must cope.

One of the first steps in establishing a consulting engineering practice is to select the desired legal entity. This decision affects many other phases of organization and operation.

chapter 10

Form of Organization

Alternatives

Three alternative forms of organization are available to the consulting engineer in the United States: an individual proprietorship, a partnership with one or more partners, and a corporation. Selection of one of these alternatives depends upon many professional, practical, and legal factors. This chapter discusses the professional and practical factors and touches upon some of the legal ones. However, for full understanding of the legal problems, the consulting engineer should consult competent legal counsel. Such assistance is needed both to aid in a decision and to prepare necessary legal documents.

As indicated in Chapter 2, 48 per cent of consulting firms are individual proprietorships, 32 per cent partnerships, and 20 per cent corporations. Initially, 60 per cent were organized as individual proprietorships, 30 per cent as partnerships, and 10 per cent as corporations.

What are the advantages and disadvantages of each form of ownership?

Individual Proprietorship

The individual proprietorship, sometimes called sole ownership, is the simplest business entity. One individual owns and operates the business and provides the required capital. All property and equipment are owned personally by him. The owner normally manages the practice although he may employ a manager. However, all contracts are made in his name and he is personally responsible. As an individual proprietor, he is personally liable for all debts, obligations, and responsibilities of the business. This personal liability extends to all assets he owns, even though they are not used in the consulting practice. The individual proprietor must pay income tax each year on the entire earnings of the business. Such earnings are added to his income from other sources and the tax is computed on the whole.

This is an easy way to begin the practice of consulting engineering. An individual simply hangs out his shingle and starts to practice. No legal procedures are required to put him into business, except registration as a professional engineer.

The advantages of an individual proprietorship for the practice of consulting engineering include the following:

1. Organization is simple and no involved legal procedures are needed.
2. Practice is legal in every state of the United States.
3. Management is direct, for the owner makes all decisions.
4. Direct personal professional responsibility is maintained.

The disadvantages of the individual proprietorship include:

1. Inability to share decisions and responsibilities.
2. Absence of continuity in case of death or extended illness of the owner.
3. Personal liability of the owner for all acts of the organization. Any or all of his assets or property may be attached for debts or judgments.
4. Difficulty of arranging deferred compensation or retirement for owner.

5. Necessity of paying taxes on full earnings annually even though earnings are not withdrawn.

There may or may not be an income tax advantage with the individual proprietorship. This subject is discussed later in the chapter.

There is a distinct place in the consulting engineering field for the individual proprietorship, which serves adequately for most one-man or small organizations. The inadequacies of the individual proprietorship become evident when there is a desire for more than one owner or when the practice grows to a substantial size.

Partnerships

Next to the individual proprietorship, the partnership is most favored by consulting engineers. This is no doubt because it allows multiple ownership while retaining the personal touch of owner-managers.

State laws regarding partnerships vary, but basic features are similar in all states. Two or more persons form a partnership, and their relationship is usually expressed in a written partnership agreement. They contribute capital, share earnings, and assign management responsibilities on any agreed basis. In most small partnerships, the management responsibilities are divided among the partners. However, management may be delegated to an employee or to one of the partners.

Earnings of the partnership are divided annually according to predetermined ratios established by the partnership agreement. Partners may receive drawing accounts against expected earnings, or, salaries that are classified as operating expenses in determining partnership earnings. Each year the partnership must file tax returns showing the profit or loss, prorated to the partners. Each partner must pay annual income tax on his share of earnings plus any salary received from the partnership, even though the earnings are not withdrawn. Each partner must add partnership earnings to his salary and other income, and compute the tax on the whole.

Partnerships have several distinct advantages, including:

1. Provision of a simple means of multiple ownership.

2. Legality to practice engineering in all states of the union, if partners are registered professional engineers.

3. Close personal professional touch of the owner-managers.

On the other hand, partnerships have certain disadvantages, including:

1. Personal liability for all debts and actions of the partnership even though incurred by acts of other partners.

2. Lack of continuity, as partnerships usually end with the death or withdrawal of a partner.

3. Liability of a partner's interest for his personal debts or obligations arising from nonpartnership activities.

4. Difficulty of reaching management decisions that normally require unanimous agreement of partners.

5. Inability to arrange key personnel ownership participation except by increasing the number of partners perhaps to an unwieldy level.

6. Difficulty of arranging for deferred compensation or retirement programs for owners.

7. Necessity of paying taxes on full earnings annually even though earnings are not withdrawn.

The relative tax advantages or disadvantages of partnerships depend upon many factors that are discussed later in the chapter.

The advantages of the partnership are such as to make it attractive to consulting firms with multiple ownership. Its disadvantages become obvious as the size of the practice grows and as financial resources and responsibilities increase.

There are ways to partially offset several of the inherent disadvantages of partnerships. Partnership agreements can be drawn to provide for continuity in case of death or withdrawal of a partner. The remaining partners may be given options to purchase the interest of the deceased or withdrawing partner. Partnership agreements may also be arranged to simplify management decisions, perhaps allowing majority decision. Insurance can be carried to lessen financial hazards, including those arising from errors or omissions.

Corporations

A corporation is the alternate to a partnership when multiple ownership is desired. A corporation is an intangible "being" endowed by law with certain powers to transact business.

Although the features of state corporate laws are not identical, the basic pattern for corporations is similar. Ownership in a corporation is represented by stock. Various kinds of stock may be issued, but usually the common stock has the voting power. Control of the corporation is vested in the stockholders, a majority of whom can influence basic decisions. Individual stockholders have no personal liability for the debts or obligations of the corporation.

The stockholders select a board of directors at annual meetings. The authority of directors varies but normally includes the election of officers to manage the corporation and the establishment of broad policies. The officers of a corporation usually include a chairman of the board, a president, one or more vice presidents, a secretary, a treasurer, and perhaps others. They are invested by the bylaws of the corporation and by the action of the board with certain administrative and executive responsibilities and authorities.

The formation of a corporation requires a charter from the state. It usually requires publication of certain notices. The organization of a corporation is a moderately complicated procedure requiring competent legal guidance.

Each year corporations pay taxes on earnings on a graduated scale. Net earnings of the corporation may be distributed to stockholders as dividends. Individual stockholders pay no personal tax on corporate earnings except as they receive dividends from the corporation. Owner-managers of small corporations usually function as officers and receive salaries. These salaries are considered as corporate expense in determining net corporate income.

Some advantages of a corporation are:

1. Avoidance of personal financial liability and responsibility for the debts and obligations of the business.

2. Continuity in case of death or retirement of an owner.

3. Multiple ownership is easily provided.

4. Management is simplified, being vested in the officers and the board of directors, which may act by majority decision.

5. Stockholders pay taxes only on dividends or salaries received.

6. Establishment of retirement or deferred payment programs for owners is simplified.

Some disadvantages of the corporation are:

1. It may not legally practice professional engineering in some states. (New York and Ohio now specifically prohibit corporate practice. In a few other states the right is doubtful, and in several states there are some restrictions.)

2. The direct personal professional responsibility of owner-manager is more difficult to maintain.

The relative income tax advantages and disadvantages of corporations depend upon many factors that will be discussed later.

The corporate form of organization offers significant advantages to large practices where there are a number of owners. Hence, in spite of precedent and opposition, there has been an increasing trend toward corporate organization in recent years, and there is every indication that this trend is continuing.

Taxes

All businesses, whether sole ownerships, partnerships, or corporations, are subject to federal income taxes. Moreover, in certain states they are subject to state income taxes.

The relative tax benefits of sole ownership, partnership, and corporation are too complicated to permit general conclusions. This is due partly to ever-changing tax statutes and laws enacted by the United States and by the various states. It is also due to variations in rates of state income taxes. Moreover, the relative income tax for different forms of organization may change appreciably at various levels of income and with various situations of personal income from other sources.

To determine relative tax advantages, a careful analysis is

needed for each situation. Frequently this calls for the help and guidance of an experienced accountant or tax expert. However, a few comments are pertinent.

With an individual proprietorship, the owner is taxed each year on the total net earnings of his business, plus his income from other sources. With a partnership, each partner is similarly taxed upon his share of the total net earnings, since the partnership pays no tax. Each partner must add partnership earnings to other income to determine tax liability. Therefore, in the case of a sole ownership or a partnership, the amount of tax is the same for a given income of the proprietor or partner.

With a corporation, however, the tax determination is far more complex because of the many alternates with respect to salaries, bonuses, and dividends. Currently the federal income tax is on a graduated percentage, increasing with total income due to a two-part tax consisting of a normal and a surcharge rate. Stockholders of a corporation, however, pay tax only upon amounts received from the corporation as salary, bonus, or dividends. A nominal dividend credit is currently permitted in computing federal income tax.

The determination of the relative tax cost with a corporation requires a careful comparison of both corporate and individual taxes under the two arrangements. United States legislation now permits a corporation to be taxed as a partnership if the number of stockholders is ten or less. Under such circumstances, the income taxes under corporate operation are the same as under partnership operation. Corporations may make such a decision only once and cannot easily go back to corporate taxation should it later appear desirable. If this option to be taxed as a partnership is not elected, corporation taxes may be minimized by paying higher salaries and bonuses to officers and stockholders. Such salaries and bonuses are considered operating expenses in determining the net profit of the corporation. Thus the tax paid by the corporation is reduced, but the owners are taxed on higher incomes. The levels of salaries and bonuses are subject to scrutiny by the Bureau of Internal Revenue and are subject to proof of their reasonableness.

Another tax advantage may be gained with a corporation if the owners establish a deferred compensation or retirement program for themselves. If this is done, the corporation treats the payments for such a program as an expense. However, the beneficiaries are not taxed until they receive the money, usually after retirement when other income is lower.

With all of the variations mentioned above, it is evident that no simple rule of thumb can indicate the relative tax differences for a corporation, as compared to a sole ownership or partnership. To repeat, it is necessary that a careful analysis be made in each situation.

Individual Proprietorship versus Corporation

So long as a practice is owned by one individual, the only choices he has are an individual proprietorship or a corporation. If he chooses the latter, he is required, in some states, to have other qualifying stockholders. This requirement is often met by having a few qualifying shares held by others, perhaps his wife, attorney, a close relative, or an employee.

The individual proprietorship has many advantages for the small practice with one owner. It avoids legal formalities, it allows practice anywhere, and it maintains personal professional responsibility. The hazard of personal liability can be minimized by adequate insurance, including an errors and omissions policy. Pressure for employee retirement programs is not great, since there is usually a small staff. Bonus or profit sharing programs can provide desired incentive compensation. The owner of a small practice, therefore, is not likely to consider corporate organization unless it offers some distinct tax advantages, perhaps related to a deferred compensation or retirement program for himself.

A corporation may also offer tax advantage if the owner desires to leave a substantial portion of profits in the business. Depending upon his incremental income tax rate, he may be ahead by reducing his personal income and paying only the corporate tax upon the profits left in the business. A corporation may also be considered if the individual proprietorship develops into a large organization where the owner must delegate considerable authority to employees. He may then desire incorporation to protect his personal investments and property not involved in the practice.

Individual Proprietorship versus Partnership

There is no reason to consider a partnership unless multiple ownership is desired. The need for multiple ownership arises from a number of sources. More than one owner may be desirable to bring technical ability into the organization, to share management and sales responsibilities, or to provide capital. Sometimes the desire for continuity prompts an established sole owner to take a younger partner who can carry on after his retirement. Once multiple ownership is selected, the organizational alternatives are partnership or corporation.

Partnership versus Corporation

From a strictly business point of view, the corporation has many advantages over a partnership. Its principal disadvantage, in the past, has been the tax situation, which, at best, has been complicated and, at worst, has resulted in excessive double taxation. Now, however, corporations with less than ten stockholders have the option to be taxed as partnerships. Thus, the tax objection to the corporation is removed for consulting engineering firms falling within this limit. For firms with more than ten owners, tax matters require careful analysis to determine relative costs. There are numerous options with respect to salaries, bonuses, dividends, and deferred compensation, which may permit a more favorable over-all tax situation with the corporation than with a partnership.

So far as management is concerned, the corporation permits majority decision and more direct assignment of responsibility and authority. The corporation gives protection against personal liability, allows continuity of the organization and simplifies deferred compensation or retirement programs for the owners.

The disadvantages of the corporation are related to professional matters. It cannot practice consulting engineering in certain states, and it makes personal professional responsibility more difficult. These disadvantages no doubt have limited the number of practices turning to corporate organizations. Nevertheless, as a practice grows, the advantages of a corporation become more apparent. Higher capital requirements increase the hazards of personal financial liability to partners and make the corporation attractive. Furthermore, in a corporate organization a larger number of employees may participate in ownership. Fifteen or twenty partners may create a difficult management situation, but the same number of stockholders can function more easily under a corporate structure.

The principal reason for incorporation, however, is the financial advantage to owners and their estates under our present high personal and inheritance tax laws. Corporate organization permits owners to take advantage of deferred compensation and retirement programs for themselves. It also helps to assure continuity of the organization beyond their death, thus making it easier for their heirs to recover the investment in the practice. Such benefits, rather than any desire to avoid personal professional responsibility, cause the present trend toward corporate structure for consulting engineers. Although this trend is most noticeable among large, mature practices, many new organizations, realizing the long-term advantages, are adopting corporate structure.

Professional Responsibility

With increased use of corporations in consulting engineering, the serious problem of maintaining professional responsibility is raised. This problem needs careful attention lest professional standards be lowered. Much of the controversy that has raged over corporate practice has centered on this problem. Most of those who object to corporate practice argue that personal professional responsibility will be lost. They raise the specter of corporations dominated by nonengineers without professional interest practicing as consulting engineers. Many of the violent opponents of corporate practice are consultants with small organizations whose needs are adequately met by individual proprietorships or partnerships. Little opposition comes from consulting engineers who, through experience with large practices, have become aware of the weaknesses of partnerships.

Nevertheless, there should be caution lest corporate practice weaken professional responsibility. It is very important for consulting engineers who choose to operate as corporations to take precautions to maintain a high professional level. It is also important for professional organizations of consulting engineers to focus attention on this need.

No responsible consulting engineer should condone practice by a corporation unless it is controlled by professional engineers and unless engineering work is done by professional engineers. One way to help assure this is to strengthen state laws relating to professional engineers. The laws of some states that permit corporate practice now require all officers to be registered professional engineers. This wise limitation should, in my opinion, be embodied in all registration laws. Moreover, I believe the majority of stock in any corporation that offers its services as consulting engineers should be held by registered professional engineers who are active in its management.

Certainly state registration laws should be modified to remove from those corporations practicing consulting engineering any exemptions from registration now permitted to business and industrial corporations. These latter organizations may not be required to use professional engineers in the design of improvements or products for their own use. No such exemption should be extended to a corporation that offers its services to the public as consulting engineers. In addition to tightening registration laws regarding corporate practice, many of the same restrictions should be included in the charter and bylaws of corporations organized to practice consulting engineering.

Personal Preference

Having had experience as a consulting engineer with each of the three forms of organization, I offer the following personal opinions regarding selection of a legal entity.

If I were the sole owner of a small or moderate sized practice, I would operate as an individual proprietor. If I were involved in multiple ownership with two or three other individuals, I would prefer a partnership until the size of the organization became such that there were substantial advantages in corporate organization.

Having reached that point, I would change to a corporate organization, with the following restrictions written into the charter and bylaws:

1. Ownership of stock be limited to professional engineers active in the organization with, however, a reasonable time for disposal of stock by the estate of a deceased stockholder.

2. Options be established permitting other stockholders to acquire the holdings of deceased or retired shareholders on a predetermined basis.

3. Officers of the corporation be limited to registered professional engineers.

4. The intent of the registration laws be respected and professional engineers be used to carry all engineering responsibilities.

If projects were to be handled in states prohibiting corporate practice, such contracts could be taken either by individual officers or by a partnership of principal officers.

With such precautions and restrictions, a corporate organization can maintain the personal professional responsibility so necessary for consulting engineers.

The execution of the varied services performed by consulting engineers involves several basic operations and functions. These affect the organization, personnel, and management of each consulting firm.

chapter 11

Operations

Basic Operations

The variety of services offered to clients by consulting engineers has been emphasized throughout Part One. Services offered by the major branches of engineering and their principal subdivisions include consultation, studies and investigations, reports, designs, supervision of construction, tests, surveys, supervision of operation, valuations, and court work. They may range from an hour's consultation to the engineering of a major project costing tens of millions of dollars.

Fortunately the execution of this great variety of engineering services rests upon a few basic professional operations, with related subprofessional and supporting activities. An understanding of these operations is a necessary prerequisite to the selection of an organizational structure.

There is more to consulting engineering than just engineering; management, sales, personnel, and supporting functions are also needed.

Management

The management function is present in every consulting firm whether it is a one-man organization or one with hundreds of employees. Naturally management problems are simpler in a small organization and grow in complexity as size increases.

In the broadest sense management, or administration, is guidance, leadership, and control of the efforts of a group of individuals with a common objective. It includes planning, organizing, directing, and controlling the activities of an organization. It requires numerous decisions concerning policy, organization, personnel, methods, plant, and finance over and beyond those related to engineering problems.

Policy includes decisions on type and size of practice, kind of clientele, and methods of operation and procedure. Organization deals both with legal entity and organizational structure. Personnel, perhaps the most important and difficult aspect of management, is concerned with recruiting, training, and administration of the people who form the organization. Methods refer to the techniques and procedures of operation and their effects upon quality and cost. Plant deals with the physical facilities and equipment required for operation. Finance is concerned with the provision of capital and the costs and economics of the practice.

Sales

All other functions are dependent upon the ability of the consulting engineer to obtain engagements. Without them, he is out of business in a hurry. Accordingly, whether recognized or not, the sales function is present in every practice.

The sales function, in its widest sense, includes all activities that build the reputation of the firm and place its name before prospective clients. Thus, every member of an organization has some share in sales; a fact that is particularly true of those who come in contact with clients and the public. In a more specific sense, however, the sales function includes the contacting of prospective clients, the preparation of brochures, literature and other materials, formal sales presentations, negotiations with clients, and related activities.

Project Administration

This function consists of the over-all administration of projects, with emphasis on client relationships. In some fields the person handling this function might be labeled an account executive. Whatever the title, the responsibility involves close communication with the client in order to interpret his needs, assure liaison between him and the consulting organization, and maintain good relations. It also includes over-all supervision to see that the project is carried out in a satisfactory and profitable manner. This function is present in every consulting practice. However, it is not always centralized but frequently is divided among a number of individuals.

Consultation

Operationally, consultation is one of the simpler engineering functions. It consists of discussion by one or more members of the organization with a client to give him professional advice about some engineering problem. It may require supporting analysis from other members of the organization, but generally only a few individuals participate. The responsibility for consultations usually falls on the more experienced and better qualified members of an organization.

Studies and Reports

One of the major engineering operations is to make studies and prepare reports.

Although the nature of analyses and the subject matter of reports may cover a wide range, the elements of the operation are similar. They include the gathering of data, followed by engineering and economic studies, investigations and analyses, cost estimates and determination of conclusions and recommendations. The completed studies and analyses are presented in written form, supplemented usually by drawings, charts, and tables.

The written report is duplicated in the required number of copies and presented to the client. This basic operation is generally carried out in the consulting engineer's office except for the gathering of information in the client's office, at the site of the work, or elsewhere. It frequently involves topographic or site surveys, together with field measurements. The work is done mostly by engineers but also requires the necessary drafting, stenographic, and duplication services.

Design

This operation is carried out in the consulting engineer's office except for outside trips to obtain required data, information, measurements, and surveys. The design operation may be divided into three major parts, preliminary design, detailed design, and preparation of plans and specifications. Often these overlap and are considered a single phase.

However, on major projects, the preliminary design phase can be separated advantageously from detailed design. It involves the determination of the major characteristics, general arrangement, and major dimensions. These data are presented by means of preliminary drawings and in written outlines. Then the preliminary designs are reviewed and approved by the client and by interested outside agencies before detailed design is undertaken.

Detailed design involves the engineering computations and decisions required to determine all sizes, dimensions, and characteristics for the project. The preparation of plans and specifications incorporates the preliminary and detailed designs and presents them by drawings and written text in a form suitable for use by contractor or manufacturer. The design function uses engineers and draftsmen, together with stenographic and clerical help. Since it involves a great amount of design, drafting, coordination, checking, review, and approval, it requires relatively greater manpower than report work.

Construction Administration

This operation is present when the consulting engineer's contract includes supervising the construction of any project. It also is present if the consulting engineer supervises the manufacture of apparatus or equipment. This operation follows the completion of plans and specifications, and carries through formal bidding or procurement, continuing until construction is complete and the project accepted.

Part of the work is carried out in the office and part in the field. In the office, the consulting engineer maintains records, reviews and approves drawings and material data submitted by the contractor, processes and approves payment estimates, handles contract amendments, interprets specifications, and deals with other related functions.

In the field, the consulting engineer places a resident engineer and inspectors on the site of the work. These men supervise the contractor's activities to assure compliance with the plans, specifications, and contract documents. They inspect the materials that go into the project, interpret the plans and specifications, and maintain certain records and documents pertinent to construction. On some types of projects, the field organization includes survey parties to lay out and stake the construction work.

Survey

Reports, design, and construction nearly always involve some survey work, the magnitude of which varies greatly with the nature of the project. It is maximum, for instance, on a highway project and minimum on a small extension to an existing factory.

Most consulting engineers who handle construction projects perform the survey operation. Some also perform extensive survey work unrelated to reports, design, or construction. Such work may include topographic, hydrographic, site or property surveys. The survey operation is done entirely in the field, except for the reduction of notes and plotting of results. It requires survey personnel, with necessary equipment and vehicles.

Tests

Test requirements vary enormously depending on the type of project and equipment involved. However, all test work involves certain similar basic procedures.

It requires the consulting engineer to go into a plant and to test the operation of equipment. It requires test apparatus, electrical, mechanical, structural, chemical, and other. It involves the operation of equipment, the observation of temperatures, pressures and other conditions, and, finally, the computation of test results and the preparation of a report. Test work is done in the field except for computations, analysis and preparation of the report.

Appearances

Expert testimony or personal appearance is required in court work, rate cases, and often in connection with other engineering activities. It involves preparation followed by appearances before a court, regulatory body, governmental agency, or other body.

Preparation for such appearances requires careful analysis of all relevant data and the preparation of reports and exhibits, a process often accompanied by conferences with the client and his attorney. Appearances are usually handled by the more experienced and better qualified members of the organization.

Supporting Functions

In addition to the basic engineering operations discussed above, a number of supporting functions are required in a consulting office. These may include:

1. Accounting, billing, and collecting.
2. Secretarial and stenographic services.
3. Filing and record keeping.
4. Receptionist service.
5. Library and reference service.
6. Communication services, including telephone and mail.

7. Transportation arrangements.
8. Maintenance and control of equipment and property.
9. Purchasing and receiving.
10. Personnel services and records.
11. Operation of electronic computers or other special machines and equipment.
12. Duplication of drawings, specifications, and reports.
13. Janitor and building maintenance service.

These supporting functions are discussed in Chapter 15.

Basis for Organization

Not all of the functions and operations listed in this chapter are needed in every consulting practice. However, most of them are found in an organization of any size. Once the required operations and functions are determined for a given practice, attention can be given to its organization, personnel, methods, and plant.

The organization of an enterprise affects its operation. Proper organization helps maintain high efficiency and good morale.

chapter 12

Organization

Structure

Organization deals with the structure of an enterprise, such as departments, sections, and subdivisions. It also covers the assignment of management functions, delegation of responsibility and authority, establishment of lines of communication, and similar matters. This chapter does not present a full discussion of the vast subject of the organization of business enterprises, but it outlines a few fundamentals, particularly as they relate to the problems of a consulting engineer.

Departmentation

The need for departmentation arises when an organization becomes too large to permit effective control of all activities by one person. The small practice with an owner assisted by one or two engineers, a draftsman, and a secretary has no need for departments. The owner-manager can directly supervise all his staff. However, add a survey party and departmentation has arrived, for its chief will obviously supervise the members of the party; or, if the practice is operated as a partnership, the need to divide functions and staff among the partners calls for departmentation.

From such simple beginnings, the need for decentralization increases with the size and complexity of an organization. Personnel must be divided into administrative units of manageable size, with clearly delegated functions and authorities. If this is not done, and the owners attempt to keep everything under their immediate control, the result is chaos.

Basic Factors

The consulting engineer who searches management literature for a ready-made structure is doomed to disappointment. There is no such thing as a standard organizational structure suitable for all situations. Moreover, most existing structures that might be used as examples are compromises between theoretical ideals and practical possibilities. However, there are a number of principles and criteria that govern organization. Among these are the following:

Workability. The final test of any organizational structure is its workability as demonstrated by use. No structure is satisfactory unless it works, and no structure that effectively functions can be far wrong.

Management. A satisfactory arrangement relieves top management from routine matters, freeing its time for over all planning, policy making, and major management responsibilities. It clearly assigns these responsibilities lest they be overlooked by owners more interested in engineering than in management.

Adequate control. A good structure provides adequate control at each supervisory level over the functions delegated to those below. This control is needed on all phases of operation, including planning and execution of work.

Span of supervision. No one individual should have so many people reporting directly to him that he cannot adequately supervise them. Most management authorities suggest this number be no more than five to seven subordinates. Such a limit is certainly valid where the subordinates are themselves supervisors and administrators. When subordinates perform more routine work, there is a higher limit to the acceptable span of supervision.

Communication. An important element in a smoothly working organization is good communication. Information and ideas must flow easily up and down the organizational structure and, in some instances, from side to side between parallel units. Organizational structure and operating procedures should facilitate communications.

Specialization. Good organization recognizes specialization by grouping similar functions together rather than scattering them through various subdivisions.

Supporting functions. Grouping of the secondary or supporting functions, even though varied in nature, often allows superior service to the rest of the organization and permits better administration and control.

Delegation. Good organization facilitates the delegation of responsibility and authority. Generally no one should have two bosses nor should a given responsibility be assigned to more than one individual. Always the delegation of authority must match the delegation of responsibility. An important element of delegation is to place responsibility for decision at the proper level; this applies to both administrative and technical decisions.

Line and staff. Good organization recognizes the merits of both line and staff responsibility. Line responsibility is best suited to operation, and staff responsibility to assistance, planning, advisory and supporting functions.

Adequate attention. A desirable organizational structure makes certain that adequate attention is given to all important functions. Each principal function should be assigned to an individual or a unit.

Coordination. Teamwork is not entirely the result of desire and effort. Organization can help by providing channels for communication and clearcut delegations that encourage and simplify coordination.

Local conditions. An appropriate organization makes it easy to recognize local conditions affecting the work and output. This may dictate a separate unit to maintain close touch with given clients or situations.

Economy. A final test of organizational structure is economy; care is needed lest division into administrative units create undue cost.

As the above factors are used to test a proposed structure, some of them may conflict with one another. The test of economy may often challenge otherwise good arrangements. Similarly, the tests of specialization and control often conflict and can only be resolved by judgment of their relative merits. This further emphasizes the important element of judgment in selecting an organizational structure. The very nature of a consulting engineering practice gives rise to several special situations requiring attention in organizational planning. Seven such situations are discussed below.

Conflict of Interest

All important executives and administrators in a consulting practice should be professional engineers. Not only does the client expect and legal requirements demand it, but such persons need professional stature to win the respect of their subordinates. The use of nonengineers in administrative positions must, therefore, be quite restricted.

A successful consulting practice requires both technical and administrative competence of a high level. Unfortunately, engineers who are well qualified technically often lack the aptitudes and interests needed for administration. Frequently the best administrators are not the best technical minds. Moreover, engineers with heavy administrative loads have difficulty maintaining an expert's knowledge of specialized fields. The organization structure can deal with this conflict by creating key positions for the experts as well as for the administrator. Such experts, free of administrative responsibilities, can concentrate on technical problems, serving as advisors or consultants within the organization. It often takes many years of practice before consulting engineers recognize the significance of this simple idea.

Many realize it only after several dismal efforts to make administrators out of experts.

Sales Function

How should consulting engineers handle the difficult sales function? In the small practice there is no problem, for owners can handle sales. But as a practice grows, the question arises as to the merits of delegating sales as a separate function or distributing it over available key personnel. Some firms have one or more employees who do nothing but sales work. Others assign it to several engineers who combine sales activities with other functions. Still others have no regular plan but choose the most available engineer as each prospect arises. However, sales activity is so vital that it warrants formal and specific assignment; otherwise, it is likely to suffer from inadequate attention.

Client Contact

Even more troublesome is contact with the client after the contract has been signed. This function may extend through feasibility studies, design, and construction covering several years. Is client contact to be a separate function or to be combined with studies, design, and construction activities? If the latter, the client may be passed along from one engineer or one department to another as work progresses. If the former, a project engineer or client liaison man must be assigned to each client.

There is much to be said in favor of recognizing client contact as an important separate function, with a project or supervising engineer assigned to each client. Such an engineer should be free to give top priority to serving the client and should not, therefore, be tied down with other functions. There is merit in combining the sales and client contact functions in what may be called a "project" or "client" department. An engineer in such a department must have a real talent for sales and human relations, as well as competence in project administration.

Administrative Units

The administrative units of an engineering practice may be divided according to branch or subdivision of engineering or according to type of project. The first pattern results in departments such as civil, electrical, and mechanical. On a complex project, work will be done in several departments. If type of project dictates organizational structure, a firm might have a power department, a highway department, and a building department. The power department, for example, would consist of electrical, mechanical, and structural engineers. Carried a step further, this type of approach may indicate a special task force on a single large complex project, with engineers from several branches.

The relative merits of branch and project type departmentation are difficult to assess. The decision often is dependent on available personnel. Frequently the branch pattern is selected to recognize a specialty, that is, heating and ventilating, while the type pattern is adopted in other parts of the same organization. Frequently a separate specification unit is also established.

Drafting Consolidation

Closely related to the "branch" or "type" question is the location of drafting personnel. With a branch approach, one drafting department serves several engineering departments. Historically this pattern has been favored. The alternative is consolidation of drafting and engineering personnel in design sections or departments, an action that eases the task of coordination and control but that may raise problems of supervision.

Report Department

The preparation of feasibility and other reports requires high caliber engineering judgment from the more experienced engineers in the organization. It further demands special skills for analysis, organization, and presentation of data in written and graphical forms. Where will responsibility for report preparation rest in a sizable organization? Some organizations combine it with design activities; some assign one engineer or a task force to prepare a given report, and others develop a report department. In this case, the report department needs access to the skilled specialists throughout the organization.

Construction Department

Is the construction and field function to be a separate one or should it be integrated with design and other activities? Here again there are several alternatives, including a separate construction department, special assignment of engineers on given projects, or the delegation of supervision of construction to project engineers. If a construction department is established, does it take over completely when a project moves into construction or does the design organization have a continuing role? These are difficult questions for which there are no easy answers. Once again the consulting engineer must decide on the basis of his organizational needs and available personnel.

Human Element

Whether for good or evil, human personality and ability are always injected into organizational problems, so that compromises are often made with ideal structures.

There are several ways in which such human elements affect organizational decisions. Perhaps the owners have attitudes or prejudices that rule out certain organization structures. Sometimes long-term, competent engineers are installed in positions demanding greater administrative skills than they possess. Rather than reduce title and position, alternative subdivisions may be created to avoid loss of face. On the other hand, extremely competent persons may assume greater responsibility than is healthy for the structure and thus warp it to accommodate this fact.

With a small organization, it is easy to accommodate these human elements, but with growth such compromise becomes hazardous. The larger the organization, the greater the merit of conforming to good management principles and resisting the temptation to undue compromise.

Case Studies

There is no such thing as a standard organizational chart for a consulting engineering practice, for we can find hundreds of arrangements. To demonstrate some of the principles of organization and to stimulate study of the subject, four different charts are presented on the following pages. Each deals with a hypothetical consulting engineering organization.

Case 1. The Jones Engineering Company, charted in the upper half of Figure 3, is owned by one person who employs a total staff of 15. This hypothetical organization is just above the one-man size. Its owner can no longer personally supervise all his people and is forced to establish subdivisions and delegate authority and responsibility to them.

His work is in two fields, suburban development and the engineering of bridges for county highway projects; each involves survey and construction. He has assigned the suburban work to engineer A, with one assistant, and the bridge work to engineer B. Engineer C supervises construction; a survey party of four men and a drafting squad of three complete the organization. A full-time secretary with a part-time assistant handles all of the supporting functions.

The organizational subdivisions follow type of project except for the drafting squad, which serves both types of work. No separation is made of the report function, and probably none is warranted; but the construction function is separated. No staff positions are needed, and the limited supporting functions are centralized with the secretary. This structure offers economy, for every supervisor is a worker as well as an administrator.

The most undesirable feature of Case 1 is the tremendous load imposed upon the owner even though the span of supervision is reasonable. He carries the sales and client contact responsibilities almost singlehandedly. As the most qualified engineer in the organization, he makes final engineering decisions. On top of this, he coordinates the work of engineers A, B, and C, the chief draftsman, and the survey party.

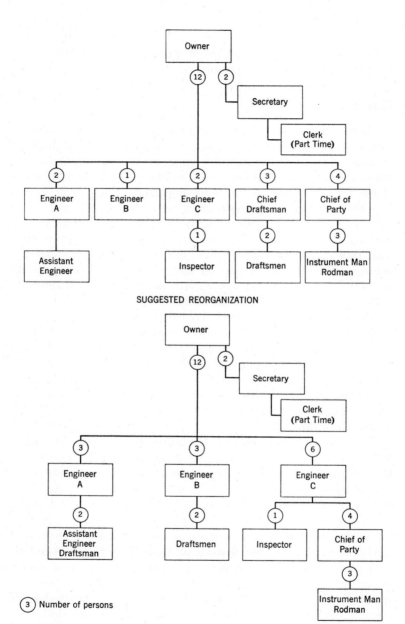

SUGGESTED REORGANIZATION

③ Number of persons

Figure 3. Organization structure of the Jones Engineering Company: Case 1, sole ownership; total staff, 15.

The owner's load can be lightened by the changes indicated in the lower part of Figure 3. Drafting personnel is divided between engineers A and B, and the survey party is assigned to engineer C who now has all outside work. Now, engineers A and B each can be delegated responsibility for the design of projects assigned to them, subject only to review and approval by the owner. Only three positions, other than the secretary, now report to the owner, and he is relieved of much of the coordination. Thus, the rearrangement gives him more time for management, sales, and client contacts.

Case 2. A, B, C Engineers, as charted in Figure 4, is owned and operated by three full-time partners. Their work falls into two major areas, sanitary engineering and structural and building work. Of their total personnel of forty, nearly half are engineers.

Senior Partner A looks after the sanitary engineering projects and supervises the business functions of the firm. Two design engineers, an electrical-mechanical engineer, and personal secretary or office manager report directly to him. He supervises the chief draftsman, who also handles work for Partner B.

Partner B looks after the building work and has two design engineers reporting to him. Partner C, the outside man, has a chief of surveys, three resident engineers, and a clerk reporting to him. The chief of surveys looks after one of the two survey parties.

Organizations like Case 2 are common among small and medium-sized partnerships. Conflict of interest between engineering and management responsibilities is present so far as Partner A is concerned. The sales function is divided among the partners as agreed from time to time. Client contact usually initiates with Partner A or Partner B, but when a project moves to construction, it passes to Partner C. Organizational subdivision is principally by type of project, but a separate drafting department is maintained. No separation is made of report function. Specialization is recognized in the position of the electrical-mechanical engineer. The supporting functions are grouped.

The workability of Case 2 depends primarily upon close coordination of the partners. Unless they have both the time and

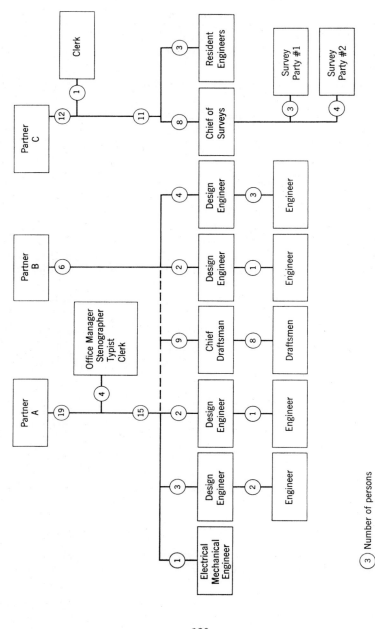

Figure 4. Organization structure of the ABC Engineers: Case 2, partnership; total staff, 40.

③ Number of persons

130

inclination to exchange information frequently, communications and coordination break down. The load on Partner A is excessive, giving him little time for top management even though his span of supervision is acceptable. Delegation of authority, control, coordination, and communication can be accomplished if the partners communicate adequately. One weak point is the dual authority over the chief draftsman. The structure fails to emphasize sales and report work adequately. The economy of the arrangement should be satisfactory, since all supervisors, including the partners, are in working positions.

The principal problems with such an organization are coordination and communication at the partner level and the absence of a direct chain of command. Important action, even of an administrative nature, requires a conference of the partners whose decisions, if there is any difference of opinion, will be a compromise. These are the hazards and weaknesses of a partnership with all partners participating on an equal organizational level. Many of these weaknesses can be overcome by the organizational structure considered in the next case.

Case 3. The Associated Engineers is also a partnership but with a substantial difference in structure. (See Figure 5.) It is a larger firm with a staff of one hundred, which designs industrial processing plants in the petro-chemical field. Although this involves chemical, electrical, mechanical, and structural engineering, all of these branches are related to the same kinds of projects.

The management arrangement, too, differs from the partnership of Case 2. By agreement the four partners have selected one as a general manager, and each of the others administers a department. The partners meet periodically to set general policies and, thus, function much as a board of directors.

Five departments have been established, namely: report, chemical, mechanical-electrical, structural, and service. Three of these—chemical, mechanical-electrical, and structural—are branch subdivisions. The report department is a functional subdivision, and the service department handles the supporting functions.

Drafting is handled differently in the various departments. Two of them have separate drafting subdivisions, while in the

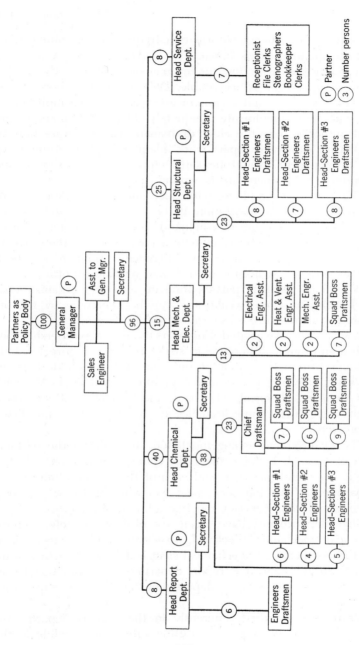

Figure 5. Associated Engineers: Case 3, partnership; total staff, 100.

third (structural), drafting is consolidated with engineering. The report department assures adequate attention to this important phase, the forerunner of design engagements. Construction activities of the firm are minor and are handled by engineers from the various design departments on part-time assignments. Client contacts are handled by engineers within the chemical department through the over-all guidance of the general manager. The organizational structure of Case 3 relieves the general manager of detailed administration and allows him to concentrate on over-all management and sales. He has two staff assistants to help him in these areas. The arrangement gives satisfactory span of supervision at all levels and is well suited to delegation, control, coordination, and communication.

Case 3 is an example of organization upon "branch lines." Its principal weakness is in coordination among the chemical, mechanical-electrical, and structural departments, all of which work on each project. Responsibility for coordination is placed with section heads in the chemical group who function as the "lead" engineers on the projects assigned to them. These same men have the basic responsibility for liaison with client. Such an arrangement, tolerable in this case, becomes burdensome if client liaison or construction supervision requires appreciable time. Then a separate construction department or other relief would be required. One other change might be the appointment of a chief engineer, reporting to the general manager, who supervises the three "branch" departments, chemical, mechanical-electrical, and structural. Furthermore, the chemical department is of such size and importance that an assistant head or a staff person is probably warranted.

Case 4. In this example, X, Y, Z Engineers, Inc. is a corporation with a large practice, 250 people, in the fields of power, civil, and industrial projects. (See Figure 6.) This company is owned by five stockholders, all active in the practice. They constitute the board of directors, serve as officers, and divide administrative responsibilities.

This organization recognizes and avoids the serious conflict of interest between client contact and design work by establishing a project group responsible for sales and client liaison. Re-

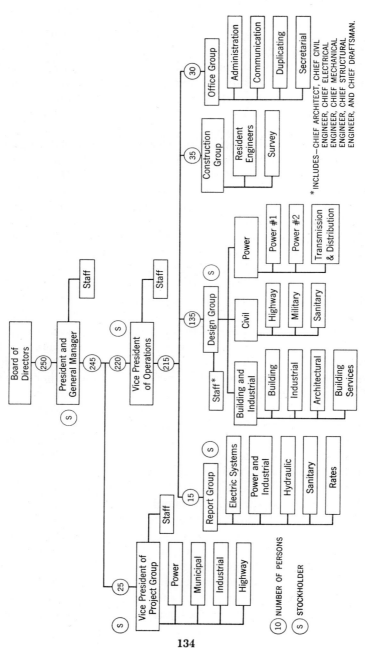

Figure 6. XYZ Engineers, Inc.: Case 4, corporation; total staff, 250.

*INCLUDES—CHIEF ARCHITECT, CHIEF CIVIL ENGINEER, CHIEF ELECTRICAL ENGINEER, CHIEF MECHANICAL ENGINEER, CHIEF STRUCTURAL ENGINEER, AND CHIEF DRAFTSMAN.

10 NUMBER OF PERSONS

S STOCKHOLDER

port and construction work is assigned to separate departments to assure adequate attention.

One vice president supervises the four operating groups and another the project group. These vice presidents and the general manager constitute top management. Each is freed from routine administration and allowed to concentrate on management activities.

Span of supervision, delegation, control, coordination, and communication are good. With the project approach in the design group, a section head has complete responsibility for a project design, except for a few specializations. The building service section and the architectural section handle these specialties for other departments and sections.

Staff assistants are utilized to assist the general manager, the vice presidents and the head of the design group. "Chief" engineers are designated, who are experts in their fields, but with no administrative responsibilities. They serve in staff positions as consultants and advisors within the organization, assisting the project, report, design, and construction groups as required. They are assigned to the design group to facilitate supervision, for most of their time is used there.

All supporting functions are placed in an office group whose operations can be coordinated with the report, design, and construction groups by the vice president of operations. The project group is closely in touch with the client and reflects local conditions and requirements.

Case 4 uses more levels of supervision and a higher percentage of people in administrative and staff positions than the previous cases. This is an inevitable trend as an organization grows in size. These added supervisors and staff must and should increase coordination and efficiency. If they don't, the arrangement becomes uneconomical and over-all costs rise.

Case 4 reflects many of the features that my firm has found desirable. The advantages of a separate project department are particularly impressive, for it allows adequate attention to sales and client contact. The project type divisions in the design group allow consolidation of engineering and drafting. Finally, there is an advantage in having separate departments for the report and construction activities and also for the supporting functions.

Comment

To repeat, there is no standard of organization that fits all consulting engineers. A structure must suit the size and type of the practice and the capabilities of the individuals available. The surest test of an organizational structure is its workability. Any structure that functions without undue tension and turmoil, and with satisfactory economy, has proven itself. Unfortunately a structure that functions today may not be satisfactory tomorrow. Changes in practice and personnel soon call for changes in structure, which should always be flexible and subject to revision.

Documentation

Once the structure has been determined, it should be documented and circulated to all employees. Four devices that can be used for such documentation, organizational charts, job descriptions, titles, and manuals of procedure, are mentioned below.

Charts. Diagrams similar to the figures in this chapter are helpful if published as organizational charts. Such charts may be of the skeleton type, as used herein, or they may be extended to show major positions and perhaps even the names of all employees.

Job descriptions. Excellent supplements to organization charts are written job descriptions outlining authorities and responsibilities of various positions. A typical job description for the power department head of the organization shown in Case 4 follows.

Department Head

Function

The Department Head of any design department shall supervise all work and personnel of the design sections and staff members included in his department.

Accountability

The Department Head of any design department reports to and is accountable to the Head of the Design Group.

Section Heads of design sections in his department and design specialists or consultants so assigned shall report directly to him. He shall supervise his department through his Section Heads.

Responsibilities

1. Technical adequacy and profitableness of all work in his department including:
   a. Maintenance of high quality of engineering in the department.
   b. Maintenance of profitable performance in the department.
   c. Supervision of preparation of design outlines unless otherwise designated for given jobs by the Head of the Design Group, and department approval thereof.
   d. Supervision of preparation of survey and field data outlines and approval thereof.
   e. Approval of engineering work and plans and specifications prepared in the department or the delegation thereof.
   f. Arrangements for consultation where desirable.
   g. Development of design standards and techniques in the department.
   h. Making or securing judgment decisions at suitable levels of experience and authority.
   i. Training, developing and inspiring personnel in the department.
   j. Technical coordination of all work in the department.
2. Administration of all affairs of the department including:
   a. Assignment of personnel within the department (Assignment of work to Sections will be made by Design Group Head after conference with Department Head).
   b. Other personnel matters within the department.
   c. Calling of necessary conferences within the department of either a general nature or on specific projects.
   d. Maintaining required design schedules and completion dates.

   e. Approval of time slips, absences, etc., or delegation thereof.

   f. Conferring with Design Group Head or his assistants concerning office space and facilities within the department.

  3. Other responsibilities as assigned by the Head of the Design Group such as:

   a. Consultation with or assistance to Project Group or others outside the Design Group.

   b. Assistance in sales work.

   c. Assistance in training programs.

   d. Consultation and investigation for the Report Section.

   e. High-level representation of the company in his field.

The Department Head shall avail himself of the assistance of staff members in handling administrative matters in order to conserve his own time. He must, however, remain responsible for all administration within his department.

Titles. Understanding of organization may be helped by using titles that are descriptive and are developed in conjunction with organization charts and job descriptions.

Manuals of procedure. Often organizational charts, job descriptions, and titles are incorporated into a more comprehensive document, called, perhaps, a "manual of procedure." Such a manual may also contain statements of normal practices and procedures (as suggested hereafter in Chapters 13, 14, 15, 16 and 17).

The preparation of materials to document organization has a twofold advantage. In the first place, it informs the organization and reduces confusion and misunderstanding. In the second, the process of preparation compels examination of many matters related to organization. This, in itself, is beneficial, for often weaknesses that need correction are discovered.

However, the publication of charts, job descriptions, and manuals is a major undertaking. They require constant review and amendment to keep pace with changes, particularly if the organization is expanding. Still, it is a worthwhile effort if it compels management to face the problems of a dynamic and changing organization.

*Because people are the all-impor-
tant ingredient in a consulting
engineering practice, personnel administration is a most vital management activity.*

chapter **13**

Personnel Management

Importance

It is sometimes said that management is concerned with five
M's: men, money, machines, methods, and materials. In a consulting practice, the first of these—men—is the predominant
factor, chiefly because the consulting engineer sells only the
time and talent of his staff. When engaged by a client, he and
his staff perform a professional service that utilizes their talents,
which consist of ability, experience, and time.

As a rule the consulting engineer's problems of money, machines (or plant), methods, and materials are minor in comparison to those of personnel. Payroll—the cost of providing
personnel—amounts to between 75 and 90 per cent of total
operating costs. Moreover, the quality of engineering service is
directly dependent upon the proficiency of the personnel comprising the organization. Finally, the all important relations with
clients are largely dependent upon the human touch. For these
reasons, this chapter on personnel management is the longest in
this book and, perhaps, the most important.

In no sense a complete treatise on personnel management, this chapter merely highlights the important elements, problems, and goals of personnel management, with particular reference to consulting engineers. It seeks to stimulate thought and interest rather than to answer all personnel questions. Anyone who manages a consulting practice will profit from an extensive study of the subject.

Primary Objectives

Personnel management is concerned with the selection and maintenance of an adequate, competent, and efficient staff to execute the engineering engagements obtained by the consulting engineer.

This staff will consist of three classes of employees. The first group, called professional, includes professional engineers, engineers-in-training, and architects in the same categories. The second group, designated here as technicians, consists of draftsmen, engineering aides, survey personnel, and other technicians who perform work related closely to engineering but at a subprofessional level. The third group, called service personnel, comprises the secretaries, clerks, and other nonprofessional staff.

To achieve competence and efficiency, personnel management must be concerned with the numerous facets that affect enthusiasm and morale, and engender a creative and professional atmosphere. These elements develop the climate in which coordination, teamwork, efficiency, and creativeness can thrive.

Special Situations

A consulting engineer takes a significant step forward in personnel management when he clearly recognizes some of his special problems, which differ from those encountered in ordinary businesses.

One such problem is the importance to engineers of professional recognition. Engineers, particularly those with a few years' experience, want to be treated as professionals. Such treatment includes recognition of their professional status and accomplishments. It also involves provision of working areas, furniture and equipment in keeping with the dignity of professionals, and a type of supervision appropriate to professionals.

A second problem is that good engineers frequently are not well qualified for administration and management. This fact is true despite their innate belief that all engineers are well qualified to be executives. Among engineers there is, unfortunately, a rather low correlation between superior technical and management aptitudes. Engineering attracts those interested in machines, devices, and material things, and inclined toward mathematics and science. But, the basic interests and aptitudes of a good manager run toward people and human relations. The wise consulting engineer will exercise great care in choosing men to be advanced to administrative positions. He will also provide alternative means of advancement for engineers who should become technical experts rather than administrators.

A third difficulty is the inherent barrier to advancement that confronts technicians and service personnel. A technician cannot be given engineering responsibility until he qualifies himself as a professional engineer by achieving registration. This is true even though he may be far more valuable than some engineers with less experience. A lesser barrier confronts service staff seeking to become technicians. Therefore, personnel policies should offer opportunities for satisfaction and achievement to technicians and service personnel within their own classifications.

A fourth peculiarity confronting the consulting engineer is the extreme dependence upon human judgment in the execution of engineering assignments. There is no easy way to measure the quality of an engineering design until the machine or the project is completed and put into service. There is no simple quality control. Even though designs and calculations are checked and reviewed, the quality of engineering is largely dependent upon experience and judgment. Accordingly, personnel policies must encourage creative and responsible professional judgment.

With these facts in mind, we are ready to examine personnel management on three fronts, policy, compensation, and practices.

Policy

Attitude

The personnel policy of any organization reflects the philosophy of its management toward the people who work for it. Personnel policy, whether expressed or written, is management's idea of how to get along with its staff.

Years ago this fact was emphasized for me by the president of a manufacturing concern that has an outstanding record of good personnel relations. He told me that other companies often sent personnel directors to his plant to see how his program worked. This president always told such visitors to return to their company and ask their chief executives to make the visit. "For," he said, "no forward and progressive personnel program will work unless it is an honest expression of the attitudes and desires of top management."

Personnel policy, therefore, starts with the owner's attitude toward employees. Are they a commodity to be hired and fired and used to advance management's aims? Or are they individuals, striving toward personal goals they hope to reach, as they assist their employer to serve his clients better and to earn a fair profit? These questions are the root of all labor-management controversy. The trend over recent decades has been toward the more liberal view that employees are human beings. The harsh, selfish policies of old have largely run their course and given way to a more humane approach.

If ever there is a field where an enlightened approach is warranted, it is that of consulting engineering. Only in a wholesome employee-employer atmosphere can one expect creative and cooperative performance from highly individualistic and professionally minded engineers.

Aims

An adequate personnel policy for a professional organization such as a consulting engineering firm should at least include the five goals listed below:

1. Give fair and impartial treatment, administered with courtesy, respect, and dignity, to every employee.

2. Pay adequate and fair compensation and, in return, obtain an honest day's work.

3. Develop and maintain a mutual professional attitude with engineering employees and accord them the maximum professional recognition.

4. Promote a sense of belonging to the organization among employees. Shape communication, information, participation, and supervision activities to encourage this.

5. Provide working conditions, associations, and procedures that are pleasant and efficient and that promote teamwork and high morale.

These basic aims, if sincerely believed and advanced and if properly implemented and administered, will lead to wholesome management-staff relations.*

Compensation

Definition

Compensation deals with both monetary reward and fringe benefits. Monetary compensation includes base pay, overtime (if paid), and incentive compensation. Fringe benefits usually cover paid holidays, vacation, sick leave, and various insurance and retirement programs.

---

* In the execution of personnel policies, consulting engineers deal directly with their employees. To date, unionism has been a negligible factor in the consulting engineering field due to the prevailing professional attitude and the small size of most firms.

Professional engineers and their societies have long contended that unionism conflicts sharply with professional principles and attitudes. Generally speaking, the only inroads have been in the engineering departments of some manufacturing concerns or large governmental agencies, principally municipal and state.

Unionism has no place in the pattern of professional engineering. It has even less place in a consulting engineering organization, where the professional attitude must be at its highest.

Base Compensation

Base compensation for professional engineers normally is stated on an annual or monthly basis, and paid monthly or semimonthly. However, some consulting engineers state base compensation for professionals on a weekly basis, and a few use an hourly basis. Base compensation for technicians and service staff is likely to be on a monthly or weekly basis, with some use of hourly rates for employees eligible for overtime. Base compensations usually cover a normal work week, predominently 40 hours but shorter in metropolitan areas and longer in rural ones.

Overtime

The consulting engineer should fully acquaint himself with the federal wage-hour regulations before establishing his policies on overtime. Recent court decisions clearly indicate that consulting engineers are subject to the provisions of the wage-hour law if engaged in interstate commerce. These court decisions interpret interstate commerce so broadly that it applies to most consulting engineers. Under the terms of the wage-hour law, overtime compensation must be paid except to exempted employees classified as professionals, administrators, or executives, and meeting prescribed standards of compensation and work assignment. In general engineers-in-training and professional engineers are exempt. Similarly, certain technicians and service staff are exempt if their work is predominantly administrative or executive. However, most technicians and service personnel are subject to the provisions of the wage-hour law. Its provisions call for time and a half for work in excess of 40 hours per week or 8 hours per day, and double time for Sunday and holidays.

Most consulting engineers consider their professional employees exempt and do not regularly pay them overtime. Some do pay it, however, because of the influence of overtime compensation on technicians and service staff.

Overtime compensation for professional engineers is hardly compatible with the professional attitude. Professional engineers are hired to do a job and should be compensated on a monthly or annual basis. No consulting engineer, however, should take advantage of absence of overtime compensation to impose upon his professional staff. Salaries should be adequate to compensate for the work done. If extensive periods of long hours are required to meet peak demands, this can be recognized either in added salary or bonus.

Today many consulting engineers ignore overtime requirements for technicians and nonprofessionals. Some of them are, no doubt, liable to legal action under the wage-hour act. No consulting engineer should ignore this act unless he has competent legal advice supporting his position.

Incentive Compensation

In addition to base salary, many consulting engineers use some form of incentive compensation for part or all of their employees. The aims of incentive compensation are to encourage greater employee interest in the firm's progress and profit, and to reward the employee for his contribution.

The principal of incentive compensation is good, but the selection and activation of a proper program are difficult. Among the alternatives are annual bonuses, profit sharing, deferred compensation, ownership, and combinations thereof. Of these, the annual bonus is the most simple and common. It is a lump sum payment made to an employee, usually at the end of the year. The amount, determined by the owners, generally reflects the profitability of the business as well as the individual's value and performance. The administration of bonuses is simple when they are small in amount and given to most of the organization. They are also easily administered when limited to a few key persons whose work is closely observed by the owners. However, if bonuses are extended to a sizable group, some of whom cannot be closely observed by the owners, administration becomes more difficult.

Profit-sharing programs are of infinite variety. They distribute some portion of profits to part or all of the employees. The amount may be determined by the judgment of the owners or computed by a predetermined formula. The distribution among employees is usually in proportion to their base earnings, but sometimes other factors are included, such as length of service or responsibility of position. Some profit-sharing programs exclude top personnel and others may limit profit sharing to income up to a certain level. Where such restrictions are imposed, another incentive is normally used for those excluded or restricted.

Deferred compensation is a procedure whereby certain sums are placed in a trust fund or insurance policy for the future benefit of certain employees. Payments to these employees may be made after retirement at a time when they will have little other income. The employee pays income tax on deferred compensation when he receives it, not when the amounts are set aside, but the firm treats the annual payments as expense. This method of compensation is particularly advantageous to the owners and key personnel at relatively high salary levels, who are concerned both with a tax saving and assurance of income after retirement. Such programs may be linked with other pension and retirement arrangements.

The opportunity, even to a limited extent, of becoming an owner, is another form of incentive compensation. If the organization is a partnership, participants must be admitted as partners. With a corporation, they may purchase stock. Often they are permitted to pay for their shares out of earnings. This form of incentive is ordinarily limited to a few key people acceptable as co-owners.

Compensation Levels

The consulting engineer has both a professional responsibility and a practical need to maintain adequate levels of compensation. Unless he does so, he will face undue turnover in staff and perhaps poor morale and low efficiency.

Compensation is supplemented by such intangibles as professional recognition, working conditions, future opportunities, and sense of participation. Nevertheless, the level of monetary compensation, including base salary and incentives, must be in line with that offered by other employers if the consulting engineer is to successfully obtain and maintain a staff.

Compensation levels for professional engineers and technicians vary in different geographical areas of the United States and between small and larger cities. These variations are related both to differences in cost of living and to demand for engineering employment. Levels of engineering compensation have risen rapidly since World War II. This has been particularly noticeable for recent engineering graduates, but, with some lag, has affected all levels of engineers. Engineering salaries tend to rise with years of experience, until a maximum level is reached. This tendency is shown in Figure 7 based on an Engineers Joint Council Report on Professional Income survey.* Such surveys are difficult to interpret because they include all graduate engineers whether they have stayed in engineering or have moved into management or businesses where incomes tend to be above those of strictly professional engineers.

Each consulting engineer must develop his own policy with respect to compensation levels. His professional responsibility is dealt with in many codes of ethics containing language similar to the following from the National Society of Professional Engineers' *Ethics for Engineers:*

> Section 21. He will uphold the principle of appropriate and adequate compensation for those engaged in engineering work, including those in subordinate capacities as being in the public interest and maintaining the standards of the profession.

Because of the constantly changing levels of engineering compensation and the geographical variations, the consulting engineer is referred to the various salary surveys that are repeated from time to time by engineering organizations.

The consulting engineer is also concerned with compensation of technicians and service staff. Technician levels of compensation are closely related to professional. In fact, most technician classifications overlap and coincide with the lesser professional categories. With respect to service employees, the consulting engineer competes primarily with other employers in the community in which he is located. Hence, compensation levels and classifications for them will be related to this market rather than to professional and technician standards.

* E.J.S. Report, "Professional Income of Engineers," 1958, p. 26.

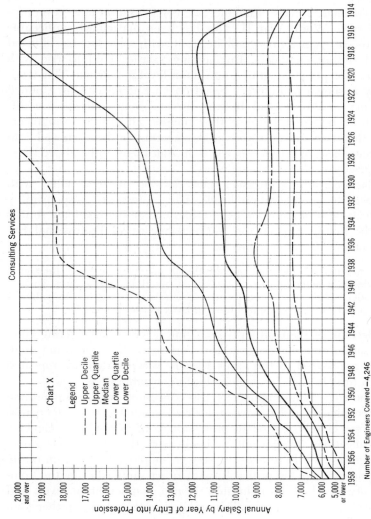

Consulting Services

Chart X

Legend
- - - - - Upper Decile
————— Upper Quartile
————— Median
- - - - - Lower Quartile
————— Lower Decile

Annual Salary by Year of Entry into Profession

20,000 and over
19,000
18,000
17,000
16,000
15,000
14,000
13,000
12,000
11,000
10,000
9,000
8,000
7,000
6,000
5,000 or lower

1958 1956 1954 1952 1950 1948 1946 1944 1942 1940 1938 1936 1934 1932 1930 1928 1926 1924 1922 1920 1918 1916 1914

Number of Engineers Covered—4,246

Figure 7. Professional salaries.

Grade Classification

The use of grades or classifications for employees has considerable merit as it brings order out of chaos in administering salaries. One such classification for engineers has been developed by the National Society of Professional Engineers, with eight grades * and is included in Appendix B. Such a classification can serve as a basis of reference for any organization as it suggests duties, responsibilities, and scope of positions; examples of work performed; typical position titles and minimum requirements of education, experience, registration, and ability. Numerous state engineering organizations have suggested minimum salaries or ranges of salaries for such classifications. A study of these is helpful in developing salary standards for any consulting organization.

Recommendations for minimum salaries for each of the eight grades vary considerably from state to state. Table 11 shows the median of seven such recommendations expressed as percentage of minimum starting salary for Grade 1:

TABLE 11. Relative Levels of Professional Salaries

Grade	Minimum Salary, %
1—Pre-professional	100
2—Pre-professional	117
3—Pre-professional	136
4—Professional	164
5—Professional	194
6—Professional	232
7—Professional	293
8—Professional	375

Classifications and salary levels for technicians are closely related to those of engineers and often overlap the lower engineering grades. One such classification for technicians is shown in Table 12.

Salary ranges for each grade, professional and technician, may

* Survey Report, "Engineers Employed by Private Practitioners," NSPE, pp. 28–29.

TABLE 12. Relative Levels of Technician Salaries

Technician Grade	Corresponding Engineering Grade	Minimum Salary, %
A	—	65
B	—	75
C	—	85
D	1	100
E	2	115
F	3	135

be substantial, often ranging from 20 per cent in the lower grades to 50 or 60 per cent in the upper ones. Frequently the upper limit for a given grade will exceed the lower limit of the next higher grade. This fact facilitates reward for experience, service, and competency when an employee is not advanced in grade, either because he lacks certain qualifications for the higher grade or because there is no opening in the organization at the higher grade.

Fringe Benefits

The term "fringe benefits" applies to a number of items of value to an employee, but that do not appear on his paycheck. Fringe benefits have become a significant percentage of base salary or wage. The consulting engineer needs policies regarding such items as:

1. *Paid holidays.* Most businesses recognize a minimum of six paid holidays: New Year's, Memorial Day, Fourth of July, Labor Day, Thanksgiving, and Christmas. Additional paid holidays may be given if it is the prevailing practice in the community.

2. *Vacations.* Mimimum policy is usually two weeks of paid vacation per year, after the first year of service. However, many firms increase vacations up to three or more weeks with long tenure. Vacation policies should cover period and scheduling of vacations, allowable accumulation from year to year, payment in lieu of vacation, minimum period of employment before vacation, and handling of vacation at termination of employment.

3. *Sick leave.* Most firms allow some paid sick leave, often one week per year, but many have no formalized program. Related items requiring decision are extent to which unused sick leave may be accumulated, minimum period of employment required for eligibility for sick leave, and availability of sick leave if not used.

4. *Compensation insurance.* Consulting engineers are governed by the various state laws with respect to compensation insurance protecting employees against injuries suffered at work.

5. *Unemployment compensation insurance.* Consulting engineers having four or more employees come under the requirements of the unemployment compensation insurance acts of the United States and of various states. They are required to pay a maximum 2.7 per cent of payroll to the state agency and 0.3 per cent to the federal agency. These rates are reduced if the firm has a favorable claim experience and may drop to only the 0.3 per cent of payroll to the federal agency. In case of layoff without cause on the part of the employee, he may claim compensation of a nominal amount for a stated period.

6. *Social security.* The Federal Old Age Benefits Program, usually referred to as "social security," is applicable to all consulting engineers having one or more employees. One half its cost is paid by the employer and the other half by the employee. Currently, the rate is 5 per cent of the first $4800 of annual earnings per individual. Both the rate and the limit on earnings have increased recently.

7. *Health insurance.* Most consulting engineers offer or encourage some form of group health, hospitalization, or medical aid insurance for their employees. These programs may be arranged through Blue Cross, which covers hospitalization and medical care, or through insurance companies, often with policies including other benefits. Some programs cover only the employee but others include his immediate family; some are paid by the owner and others are contributory.

8. *Life insurance.* Some consultants arrange a group life insurance for their employees, often at attractive rates. Such programs may be paid by the employer or may be contributory, with the employee paying a portion of the cost.

Retirement

The majority of consulting engineers do not have retirement programs due to their relatively small size, youthful age, and variable number of employees. Nevertheless, many firms have established such programs and more will find it desirable to do so as their organizations grow and mature. Such programs will be increasingly necessary if consulting engineers are to compete successfully with other employers of professional engineers and technicians. They are important because they add an element of security that helps to attract and hold a competent staff.

The subject of retirement and pension programs is an extensive one, far beyond the scope of this discussion. The development of such a program requires extensive study and competent advice to select the most suitable plan.

Practices

Administration

Numerous practices and procedures are required to administer and supervise the personnel of a consulting engineering organization. In the following pages suggestions are presented for the following: recruitment, induction, supervision, review and counseling, promotion, advancement and professional development, registration, communication, termination, and records.

Recruitment

Effective recruiting of personnel is important in any organization, but it is doubly so in an engineering firm where the chief product is knowledge and judgment.

The first step in good recruiting is advanced planning of personnel needs. When the consulting engineer, engrossed in executing the jobs on hand, fails to anticipate his coming personnel needs, he is forced to meet peak loads by the quick hiring of the first man available.

But if needs are anticipated, a number of candidates can be considered, which results in the greater possibility of a wise choice. Some personnel managers say that one should never fill a position without interviewing at least six candidates. Candidates for employment may be located by advertisements in technical journals, references from employment agencies, interviews of forthcoming graduates at engineering colleges, and referrals from college placement offices. Before applicants are sought, it is well to prepare a job description or personnel specification for the position. Such a statement should include the education and experience required and any age or salary limitations.

As applications are received, a quick screening will eliminate the unqualified and select those for further investigation. For a position of any importance, this should include a check of references and a personal interview, preferably at the engineer's office. Some consultants use psychological tests recommended by many personnel authorities. I have found these helpful, not to determine technical qualifications, but as a guide to potential growth as a specialist or an administrator and as a warning of instability or potential personal problems.

Induction

The process of inducting an individual into an organization is almost as important as the selection of the right person. In a small organization, this process may consist of a short talk with the owner who has hired him, after which he is turned over to his immediate supervisor who completes the task. In larger organizations, a more formal orientation program is helpful, perhaps extending over several weeks.

The new employee should understand all personnel policies, including the fringe benefits for which he is eligible, the work and activities of the organization, and the procedures that he must follow. So far as his immediate assignment is concerned, he should be thoroughly appraised of his authorities and responsibilities and of relationships to superiors and subordinates. To assist in orientation, many consulting engineers publish a personnel manual containing pertinent data.

Supervision

Supervision of personnel in a consulting engineering firm involves all the problems and principles usually associated with the guidance of employees. The problem is to have adequate supervision to assure effective use of manpower, but not so much as to hinder initiative and individual development. Supervision involves assignment of duty and delegation of authority, followed by adequate control. Probably no better statement of basic principles of supervision has ever been written than that prepared by the American Management Association, which is reproduced below:

1. Definite and clean-cut responsibilities should be assigned to each executive.

2. Responsibility should always be coupled with corresponding authority.

3. No change should be made in the scope or responsibilities of a position without a definite understanding to that effect on the part of all persons concerned.

4. No executive or employee, occupying a single position in the organization, should be subject to definite orders from more than one source.

5. Orders should never be given to subordinates over the head of a responsible executive. Rather than do this, the officer in question should be supplanted.

6. Criticisms of subordinates should, whenever possible, be made privately, and in no case should a subordinate be criticized in the presence of executives or employees of equal or lower rank.

7. No dispute or difference between executives or employees as to authority or responsibilities should be considered too trivial for prompt and careful adjudication.

8. Promotions, wage changes, and disciplinary action should always be approved by the executive immediately superior to the one directly responsible.

9. No executive or employee should ever be required, or expected, to be at the same time an assistant to, and critic of, another.

10. Any executive whose work is subject to regular inspection should, whenever practicable, be given the assistance and facilities necessary to enable him to maintain an independent check of the quality of his work.

An added caution to consulting engineers is to respect the desire of engineers to assume professional responsibility and carry out assignments with a minimum of close supervision.

Review and Counseling

A key element in personnel management is the periodic review of an employee's performance, accompanied by effective counseling. Every employee is entitled to a periodic appraisal of his work by management. The old adage that no news is good news hardly meets the situation, but many supervisors fail to comment on a subordinate's work except when unsatisfactory.

Day-to-day communication between a supervisor and his subordinates should be supplemented by more formal counseling. The wise supervisor will create opportunities for personal and private visits with his staff about their work, progress, opportunities, and aspirations. Periodic employee appraisals form an excellent base for counseling with individuals. Counseling is an important function that requires considerable skill and practice. It should be done by individuals thoroughly alert to human relations and in a position to reflect management's policies and plans. The dividends from good counseling can be substantial, but the results of poor counseling can be embarrassing and damaging.

From the management point of view, employees should be appraised and reviewed periodically. Such action is necessary when considering salary adjustments and promotions. It is also helpful in weighing and guiding each employee's development, in spotting those of outstanding potential and as a basis for counseling.

An annual appraisal of each employee is a minimum requirement, but more frequent reviews are warranted for new employees and for those in the lesser technician and service categories. Such appraisals may be quite informal in a small organization; but, with a larger concern, they should involve a written appraisal by the individual's superior and perhaps others. Rating or appraisal forms are helpful. Special appraisals may be appropriate when considering individuals for promotions or when dealing with personnel problems.

Promotion and Advancement

Individuals may be advanced in salary two ways, first, by merit adjustments fitted to growth in experience and ability in a given position and, second, by promotion to a position of greater responsibility.

Merit adjustments are best dealt with by periodic review of all staff, perhaps once a year except for the lower grades where semiannual review is desirable. If an organization has established job descriptions, job classifications, and salary ranges, merit adjustment for a given position is simplified.

For an individual to advance to a higher position, two conditions must be present. There must first be a vacancy resulting from the promotion or termination of one previously holding the post or from the creation of a new position. Next, management must be convinced the individual is qualified for the post. Whenever management is faced with a vacancy, the question arises as to whether it will be filled from within or without. Promotion from within is normal, provided suitable candidates are available. Higher morale is maintained if employees know they will be given first consideration when an opening occurs. Nevertheless, there are times when no one within the organization is ready for the advancement, and management has no alternative but to bring in new personnel. If management frequently finds it necessary to go outside to fill positions of responsibility, it is an indication of an inferior job of personnel selection, training, or evaluation. Appraisal and evaluation of employees' growth potential is difficult but exceedingly important. There is some truth in the saying that familiarity breeds contempt. It is easy to overlook talent and pick someone from outside, only to find him no better than employees who were passed over.

Before a selection is made, the required qualifications of the position should be carefully matched against the aptitudes, characteristics, and abilities of the candidates for the position. As this is done, the consulting engineer will be able to appraise the relative demands of administrative and technical proficiency. It is my observation that consulting engineers often err by failure to distinguish between these last two characteristics. It is easy to advance the competent and dependable engineer to an administrative position for which he has neither the aptitude nor the interest. If management will distinguish between these technical and administrative talents, dual paths may be provided for engineers to develop and grow within the organization.

Professional Development

Professional development is largely an individual responsibility, but the wise consulting engineer can do many things to encourage it among his employees. Professional development includes a range of items. Some years ago, I wrote the following definition: "Professional Development is the process by which an inexperienced engineering graduate transforms himself into a useful and active citizen as well as a competent professional engineer," and suggested a test for measuring professional development. *

Several activities and programs that will advance professional development are:

1. Encouragement of registration as professional engineers.
2. Encouragement of membership in professional and technical engineering societies and attendance and participation at engineering group meetings.
3. Encouragement of the writing of technical papers and articles.
4. Encouragement of activity in community, civic, church and other nonprofessional activities.
5. Conduct of seminars and discussion groups within the organization.
6. Support of evening or Saturday classes for advanced credit.

Since actions speak louder than words, one of the greatest contributions to the professional development of young engineers is the example of outstanding professional conduct and interest by the top men of a firm.

* "How is Your P.D.?" by C. M. Stanley, *Mechanical Engineering*, August 1950, p. 692.

Registration

The mark of the professional engineer is registration by one or more state boards of engineering examiners. Since registration is an important element of professional recognition, the consulting engineer needs a definite policy toward it. Beyond general encouragement, he can promote registration by using it as a factor in determining salaries, assignments, and advancements. All positions above a certain level should carry registration as a prerequisite. Similarly, salary advancement beyond a predetermined level can be conditional upon registration. Such policies will accelerate registration of young graduates as engineers-in-training and encourage rapid achievement of registration as professional engineers. They will result in a high percentage of registered engineers.

Communication

No single problem involved in personal management is more thorny than that of communications. The term as used here applies more to the flow of general information and ideas through an organization than to the flow of instructions and orders. The difficulties of maintaining satisfactory communications are matched only by their importance.

Communication, although simple in a small organization, obviously becomes more complex in a larger one. Often the heads of an organization lose personal touch with all but a few of their employees as it becomes necessary to departmentalize. Good communications then become a necessary substitute for the closer personal contact that is possible in a smaller organization.

Theoretically such communications are possible through organizational channels. Information conveyed by the owners to their immediate subordinates should find its way down to all members of the organization. Practically, however, there are too many opportunities for information to become lost or distorted, and there are too many blocks against upward flow of ideas. Nevertheless, organizational channels should be used and every effort exerted to increase their effectiveness.

Such channels of communications may be supplemented by personnel counseling. To be effective, the authority and responsibility for it must be specifically delegated to supervisors who are trained and schooled in this process. Another good method of communication is group contacts between owners and staff. These will be more effective if groups are kept small and if free discussion is possible. These personal communications devices may be supplemented by written communications, including a house organ, memoranda, and information. These can be helpful in informing members of the organization of the owners' ideas and in familiarizing them with the problems and progress of the organization.

Termination

Terminations may be instituted either by an employee or an employer. In the first case, the employer has little to do except arrange the time and conditions of departure. A terminal interview, however, is desirable to fully ascertain the reasons for the termination. Such an interview, preferably conducted by an owner or a principal, may be valuable in finding and correcting unsatisfactory conditions. A terminal interview may also create a wholesome feeling on the part of both the departing employee and the firm.

Where terminations are activated by the employer, however, he has a substantial responsibility. First, the decision to terminate, though difficult and sometimes unpleasant, should be made forthrightly. Procrastination and delay solve nothing. When an employee is released by the owner, he should be told the reasons for his dismissal. I have always believed this should be done personally by a principal or owner. Every effort should be made to help the employee to improve himself in the future. On the other hand, when terminations result from lack of work, the decision is less difficult and may well be anticipated by the employee. Even so, he is entitled to reasonable notice and to final review or counseling with a principal.

Records

Good personnel records are a fundamental of personnel administration. They should include:

1. Vital statistics respecting the employee and his dependents.
2. Qualifications prior to employment including education and experience.
3. Assignments and positions held during his employment.
4. Salary levels and annual compensation.
5. Reports on performance ratings, tests, and counseling.

Along with major decisions on operations, legal entity, organizational structure, and personnel management, a multitude of important details affect procedure. For these, the consulting engineer must establish methods and techniques.

chapter **14**

Methods and Techniques

Scope

The methods and techniques discussed in this chapter are concerned with the handling of work in an engineering office, not with engineering itself. They facilitate engineering, record the results of engineering studies and analyses, and aid communications.

The engineering of sizable projects involves the handling of a mountain of paper: computations, letters, drawings, plans, specifications, contracts, payment estimates, and reports. Standard procedures avoid confusion, minimize costs, and avoid wasted effort.

The methods and techniques discussed below are often illustrated by practices adopted by the author's firm. Obviously there are numerous alternatives, some of which may be superior to those presented. It is more important, however, to use a workable method consistently than to strive eternally for perfection.

Identification

A consulting engineering office might be likened to a job shop manufacturing plant, which simultaneously works on a number of specific engagements, each tailor-made to the requirements of the client. A means of identifying these engagements is required for accounting, filing, scheduling, and related functions.

A common practice is to use a job or order number that is applied wherever identification is needed. A simple system uses serial designation, with numbers assigned in order as engagements are received, that is:

3846—City of Waterloo—Sewage Treatment Study
3847—Iowa State Highway Commission—Highway 34
3848—Jones Manufacturing Company—Plant Extension

One of the alternative systems uses a two-part number; the first portion identifies the client and the second, the number of the engagement for that client, that is:

186-4—City of Waterloo—Sewage Treatment Plant

It makes little difference what method is adopted so long as it is followed consistently. Years ago, our firm started the serial designation. Were we to do it today, we would use the two-part digit suggested above.

Job Orders

Some form of authorization is desirable to "open" a new job and disseminate information about it. Data that may be useful include: job number, data, client's name and address, nature of engagement, brief description of work to be done, fees, schedule, assignment of key personnel, and information on related engagements. Such information may be transmitted in a memorandum, but it is simpler to use a job order form with multiple copies. These go to each man having a key responsibility on the engagement and to heads of the units handling the work. A copy should be circulated to such places as the file room and accounting department. Similarly, a job closing order is helpful to assure that all proper steps are taken when a job is finished. It activates preparation of final billing, completion of records and files, and other steps associated with the closing of a job.

For the record, jobs should be listed both alphabetically by clients and serially by numbers. For current use, it is helpful to prepare "job lists" of active engagements, including job number, client and project designation, and perhaps key personnel assignments.

Approvals

One sound engineering practice is that work must be adequately checked, reviewed, and approved before it is released. This practice guards against error and minimizes the chance of poor judgment. However, to assure adequate checking and review, it is necessary to establish definite procedures and to provide a means of recording such steps. The latter requirement is easily met by affixing the initials (or name) of the individual performing the operation to the computation sheet, drawing, draft, or report in question. To facilitate this, such documents are often provided with spaces for initials of the originator, the checker, the reviewer, and one or more approvers.

It is difficult to determine how many checks, reviews, and approvals should be required. Any drawings or analyses of importance should be checked by a second engineer to guard against errors in arithmetic, method, and judgment. Often it is desirable to add a "review" step by a more experienced engineer not involved in the initial execution. Beyond this, over-all approvals are usually required of completed sets of specifications, drawings, or reports. Sometimes dual approvals are desirable, depending upon the organizational structure.

The maintenance of adequate standards for checking, review, and approval are important elements in sustaining high quality. Equally important is the choice of the individuals to review, for their approval is no more valuable than their experience and ability.

Computations

Numerous computations are required for analysis and design in a consulting office. Standardized practices are desirable to obtain uniformity in the preparation, filing, and indexing of these computations. This not only assures higher quality but simplifies future reference.

COMPUTATION SHEET

Form No. 15

STANLEY ENGINEERING COMPANY — Job No. ________ Page No. ______

Computed by______ Date ______ — Subject ____________________

Checked by ______ Date ______

Reviewed by______ Date ______

Approved by______ Date ______ — Sheet No. ______ of ____________

TITLE CORNERS

WESTERN POWER CORPORATION

REVISIONS						
NO.	DATE	ITEM	DWN	CKD	RVD	APP
1	5-6-58	Stack Elev.	R.L.H.	WCB	TDW	REB

POWER PLANT EXTENSION

ARCHITECTURAL ELEVATIONS

STANLEY ENGINEERING COMPANY

MUSCATINE, IOWA — CHICAGO, ILLINOIS

DRAWN R.M.M. — DATE March 12, 1958 — NO. 3021-63 — REV. 1

CHECKED R.A.G. — SCALE ½" = 1'-0"

REVIEWED T.D.W.

APPROVED V.E.L.

WESTERN POWER CORPORATION

REVISIONS						
NO	DATE	ITEM	DWN	CKD	RVD	APP
1	1-3-58	General	JB	RG	WL	BS

POWER PLANT EXTENSION

LOT LAYOUT

STANLEY ENGINEERING COMPANY

MUSCATINE, IOWA — CHICAGO, ILLINOIS

DRAWN RLH — DATE 7-24-57 — NO. 3021-2 — REV. 1

CHECKED RD — SCALE 1" = 80'-0"

REVIEWED TDW

APPROVED REB

WESTERN POWER CORPORATION
POWER STUDY

ELECTRIC DEMAND

STANLEY ENGINEERING COMPANY

MUSCATINE, IOWA — CHICAGO, ILLINOIS

DWN. K.D.D. CKD. D.C.B. APP. D.H.A. DATE 3-21-59

Figure 8. Example of a computation sheet.

To dignify computations, many consulting engineers use special computation paper, which usually has a heading including the firm's name and spaces for job number, description, date, and initials of those making it. The form of computation paper long used by the writer's firm is indicated in Figure 8.

Every office should establish standards for the form and content of computations. These should include the sources of data and references, a clear designation of assumptions, the identification of the computed answer, and the factors considered in reaching a conclusion. Such "tracks" are most helpful to those who check, review, and approve the analysis. During the process of analysis, it is convenient to keep computations in notebooks; but, when completed, they should be placed in binders and permanently indexed and filed.

Drawings

Much of the output of a consulting engineering office consists of drawings or plans. Hence the need for standardization of drawing sizes, title corners, media, and drawing numbers is apparent. Almost an infinite number of alternatives are available in this area. Desired sizes of drawings may include some of the following: 8½″ x 11″, 11″ x 17″, 18″ x 24″, 24″ x 36″, and 28″ x 42″.

With a few exceptions, the author's firm has found the above sizes satisfactory. Sometimes larger drawings are needed for maps or assembly drawings for machinery. Some of the larger sizes used are: 30″ x 36″ and 30″ x 48″.

It is usually economical to purchase the drawing media printed with margins and title corners. Content and arrangement of title corners vary with the preferences of the consultant, and the size and type of drawing. Generally, they contain the drawing title and number, the client's name and the project designation, the engineer's name and city, and the date. Often they include spaces for initials of the draftsman, checker, and approver. When drawings are used for construction purposes, title corners need space for listing of revisions. Drawings used in reports often have a simplified title corner. Several typical title corners are shown in Figure 8.

Each office standardizes on the media used for drawings. The most common is pencil on vellum, of which there are a number of varieties. The older media of ink on tracing cloth is seldom used unless required by the client. Ink on vellum is frequently used.

Drawing Numbers

Several types of drawings must be numbered and identified, including contract drawings, figures for reports, study drawings, and foreign drawings. Contract drawings are the plans for construction projects or equipment. These are preferably given drawing numbers that show the job identity; for example, a number 1824–19 indicates drawing 19 on Job 1824. Drawings and charts in reports are usually identified as figures or exhibits, that is: "Figure 5" or "Exhibit B." However, for filing purposes they, too, should carry job identification, perhaps in the margin. Some offices use a letter suffix or prefix to separate report drawings from contract drawings. If this is done, a report figure might carry the number of 248-R-21, indicating a "report" drawing on Job 248.

In addition to drawings used in reports or contracts, the consulting engineer frequently prepares drawings as studies or interpretations. If drawings are identified by jobs, a letter suffix may be helpful (the number 246-X21 representing such a special drawing). Finally, a consulting engineer must process and file numerous drawings originating outside his office, drawings from client, contractor, or manufacturer and sometimes referred to as "foreign" drawings. To facilitate filing and identification, these may be numbered with the suffix "F." A foreign drawing might be identified as 102-6-F41. The imagination of each consulting engineer will suggest numerous alternatives. The important thing is to adopt a simple, consistent system.

Drafting Standards

There is a real advantage in developing uniform standards to cover a multitude of techniques involved in drafting, such as symbols, dimensioning, arrangement, lettering, and so forth.

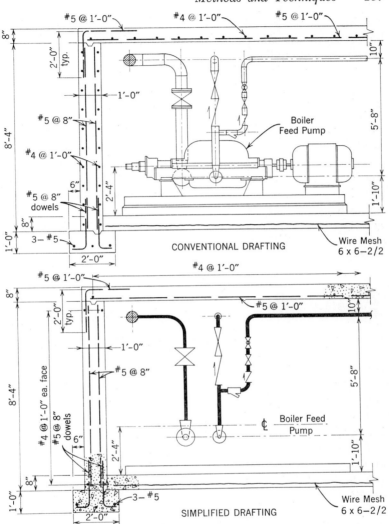

Figure 9. Comparison of ordinary and simplified drafting.

The objective of drafting is to prepare drawings that convey visual information, not a pretty picture. Occasionally a pictorial presentation is desired for a report, an architectural rendering, and such; but the purpose of most drawings is to communicate information, not to make an impression. A great amount of time can be wasted on extravagant drafting. Recently the tendency has been to use simplified drafting, which reduces the detail shown. Such simplified drafting reduces the cost of drafting and lessens the task of review and checking. A comparison between ordinary and simplified drafting is indicated on Figure 9. Simplified drafting is to be recommended.

Consulting engineers who handle repetitive work must often show the same construction detail time after time. This suggests the preparation of standard drawings (perhaps 8½″ x 11″ in size) for incorporation in specifications rather than on plans. The areas in which such standards are advantageous include: pipe hangers, standard manholes, pipe trench covers, and electrical distribution pole heads. Numerous other opportunities suggest themselves.

Design Standards

Engineering standards are just as useful and timesaving as drafting standards. These may be prepared in numerous areas of engineering and may be of the following types:

1. Charts or tables that provide solutions to repetitive, simple design problems, allowing a saving in time and an increase in accuracy.

2. Outlines of stresses, loadings, and other design criteria, which assure uniformity and save time.

3. Outlines of design approach on more complicated problems assuring a more uniform and perhaps superior analysis.

4. Check lists or procedures that minimize chance of oversight of some design matter or detail.

5. Statement of special conditions required by certain clients or situations.

Specifications

Each consultant tends to develop routine procedures with respect to the preparation of specifications. These include: methods of preparation and review, normal division of specifications into parts, and standard conditions and forms for inclusion in specifications.

Part 6—Concrete

1. *Cement*
   a. Regular Portland cement: ASTM C150, Type I; one brand for entire job.
   b. White: non-staining "Medusa White" by Medusa Portland Cement Company.
   c. Calcium-aluminate: "Lumnite" by Universal Atlas Cement Co., or equal.
   d. High early strength: ASTM C150, Type III.

2. *Regular Aggregate*
   a. Strong, durable, uniformly graded mineral grains conforming with ASTM C33 with limitations shown hereinafter.
   b. Permissible weighted loss for aggregate subjected to 5 alterations of magnesium sulphate soundness test:

Fine aggregate:	15%
Coarse aggregate:	18%

   c. Permissible limits for deleterious substances in percent by weight:

	Fine	Course
Soft fragments:	—	5.00
Clay lumps	1.00	.25
Coal and lignite:	.25	.25
Material finer than No. 200 sieve:	3.00	1.00
Other deleterious substances:	2.00	3.00
Combination of above items:	4.00	5.00

   d. Fine aggregate grading in percent by weight passing:

Sieve Size	No. 4–100 Size Aggregate
3/8"	100
No. 4	95–100
No. 16	45–80
No. 50	10–30
No. 100	2–10

   e. Fine aggregate fineness modulus: not less than 2.5; not more than 3.

Figure 10. Outline type of specifications.

Some consulting firms develop standard specifications for repetitive items, for instance, those covering concrete or high pressure piping that may be inserted in a subsequent set of specifications. Any necessary modifications to such a standard are stated in an addendum.

The difficulty with standard specifications is that variation is the rule, not the exception. Accordingly, it may be more desirable to use check lists or "sample" specifications that are altered on each project to suit the circumstances. These "sample" specifications may be reproduced in quantity for internal use and editing.

There are two ways to prepare specifications, complete text and guide specifications. The first, written with complete sentences and paragraphs, is the most commonly used. "Outline" specifications, on the other hand, state requirements in outline form and omit complete sentences and paragraphs. A sample outline specification is shown on Figure 10. The use of outline specifications has the advantages of simplicity, ease of checking, and great reduction in volume of words. Moreover, they are easier to use in the field.

Specification procedures should include a careful checking of specifications against plans to assure coordination and to reduce errors and omissions.

Reports

Engineering reports are the least uniform and standardized of all engineering documents. This is due both to the variety of content and the preferences and prejudices of various consultants. Nevertheless, there is merit in developing a definite pattern for reports. It simplifies preparation and offers a "hallmark" of identity for the consulting engineer. To develop a pattern, attention must be given to both substance and form. The objective should be an intelligible and easily read document that conveys the engineer's thought and recommendations. Since many reports are for laymen, the engineer should avoid technicalities and write in terms easily understandable to the nonengineer.

The consulting engineer might consider the following items in developing methods and techniques on reports.

Substance: Presentation of results and conclusion versus presentation of study itself.

Style of presentation and writing.

Amount of detail to include.

Arrangement: Division into parts and sections.

Use of figures versus text.

Use of appendixes.

Indexing.

Use of synopsis and summaries.

Form: Method and media of reproduction.

Type of report covers.

Format of charts, sketches, and other drawings.

Procedures for drafting, checking, and approving.

Communication and Coordination

So far this chapter has dealt primarily with form and paper. These are important, but of greater significance are the procedures that create effective teamwork and full understanding among those working on a particular project. This takes us into the area of internal communications and coordination. The consulting engineer must give personal attention to the development and enforcement of procedures in these areas, or he is apt to find his staff moving in several directions, with a resultant loss of time, effort, and money.

Procedures in this area are vitally affected by the organization structure adopted. (See Chapter 12.) Hence they require study and adaptation to each specific office. However, a few fundamentals that are necessary and several devices that are helpful are outlined below:

Lead engineer. On every assignment, report, or design, one engineer should take the "lead." He carries the responsibility for coordinating the work of the other engineers and draftsmen. The lead engineer should be intimately involved in the work. He must be present most of the time, for coordination is an hour by hour and day by day task. Thus, the lead engineer can not normally carry the responsibility for client contact since this would take him from the office. The position and title of a lead engineer will depend on the organization structure.

Requirements. A full, written statement of the job requirements and of special conditions will save confusion and misunderstanding. Such a document, sometimes called a "job outline," is often prepared at an early stage. The client-contact man should assist in its preparation to make certain it reflects the client's needs and desires. Once prepared, it can be circulated to all who have responsibility for the job.

Communication. Coordination is enhanced by good communication to inform the proper persons of decisions, changes, progress, and problems. Both vertical and horizontal flow of information is needed.

Conferences. No single device facilitates coordination more than a conference of those concerned. Frequency and participation must depend on the judgment of the lead engineer who should also determine agenda and length of the conference. Judiciously used, the conference can be a valuable tool.

Control. To assist the lead engineer a system of simple, frequent and effective progress reports are needed. These may be verbal or written, short or long. But they are essential to inform the lead engineer of situations needing control and coordination.

Final check. To assure the coordination of completed work, a final review or check by one competent to deal with all phases is helpful. This may be handled by the lead engineer himself, a checker, or other engineer.

Manuals

As standard techniques and procedures are developed within an organization, they can be documented and published as manuals. This assures uniformity and helps to orient and inform new employees. Each consulting engineer can adopt the pattern he will use for preparation, approval, release, and filing of such standards. Illustrative of the scope of manuals that can be useful is the following classification of those used by the author's firm.

Administrative or general memoranda of a permanent nature dealing with organizational structure, job descriptions, and company-wide policies and procedures.

Employee manuals that present information regarding working hours, vacations, fringe benefits, and other matters related to employee-employer relations.

Department manuals on procedures and practices within a specific subdivision of the organization, that is, design department or report department. Subjects include both administrative and technical matters (such as design or drafting standards).

Assignment and informational memos that communicate news or orders of current or temporary nature.

Manuals are effective, however, only if they are studied and used.

The efficiency of any organization is dependent upon smooth running service in the background. Arrangements for these supporting functions are important in any consulting practice.

chapter 15

Supporting Functions

Importance

In Chapter 11 various supporting functions required in a consulting office were mentioned, including: accounting, stenographic and secretarial, filing, receptionist, library, communications, transportation, property, purchasing, personnel, computer, duplicating, and building maintenance.

Most of these functions are present in every consulting practice large or small. In the small organization they may be handled by one competent all-purpose secretary. However, in larger offices, the size and complexity of these service functions suggest a separate department supervised by a competent office manager. It is important that these functions be properly organized and smoothly performed. This is true not only because they improve over-all efficiency but also because they affect public relations.

As salaries of professional engineers and technicians are comparatively high, economy requires that they be relieved of nontechnical duties. Accordingly, the objective of these functions should be competent service that saves time for the professionals and makes their work more effective.

First impressions are often lasting ones, and such first impressions frequently are made by supporting personnel. A cheerfully efficient receptionist or the pleasant voice of the telephone operator can set the stage for a worthwhile contact. The quality of letters and other typed documents carries impressions to distant places. In such ways, service functions affect the public relations of a firm.

Each of the supporting functions is discussed below except accounting, which is the subject of Chapter 17.

Stenographic and Secretarial

The stenographic load in a consulting engineering office is heavy and exacting. It includes not only the usual correspondence but also numerous and bulky reports and specifications. Technical terms, numerical values, and many tables make competent stenographers of vital importance. On the other hand, the secretarial load in most consultants' offices is less than that encountered in other offices. This is so because most of the staff are engaged with engineering problems and the number of callers is not large. As a result, many consulting engineers do not appreciate the benefits of effective use of secretaries.

In the small organization, the secretarial and stenographic duties are combined. One or two individuals may handle them, along with the duties of the receptionist, the telephone operator, and the file clerk. However, in larger organizations, secretarial duties may be separated from stenographic ones. Frequently the work load in the various parts of an engineering office fluctuates so much that a stenographic pool is desirable. The relative advantages of a pool versus individually assigned secretaries or stenographers is a constantly recurring problem. It is related to the extent secretaries are assigned to principals and is somewhat dependent upon the use of dictating equipment. If a number of secretaries and stenographers are required, they should be under the supervision of a chief secretary or an office manager. This person can assume the responsibility for maintaining an adequate staff, training, supervising, establishing standards, and assigning work at rush times.

Files

Files are required for general correspondence, drawings and tracings, blueprints, plans and specifications, field notebooks, and prints of foreign drawings.

The satisfactory operation of filing departments requires selection of equipment, determination of methods and procedures, and assignment of competent personnel. With a sizable organization, files should be isolated under the charge of responsible file clerks to whom the authority to remove and replace materials is strictly limited.

The following questions must be decided: Will there be a central filing system or will files be distributed? To what extent will duplicate files be maintained? What method of indexing will be used? What procedures will be established for the check out and return of materials? How long will materials be maintained in the files? Are there to be both active and dead or transfer files? Is microfilming justified to reduce the bulk of stored material?

Receptionist

One of the most important people in any organization is the receptionist, who makes the first impression on a newcomer. She should be alert, pleasant, and courteous. She should know the organization and how it functions. When salesmen, contractors, and suppliers call regarding some project or product, she must direct them to the proper party. The receptionist's task is further complicated when she is also responsible for operating the telephone switchboard.

Communications

Communications for a consultant include telephone (internal and external), postal and telegraph services, and internal messenger service.

An office with more than a handful of extensions needs a switchboard, manned by a trained operator, often the receptionist. In larger offices an automatic or PBX system is indicated. In either case, internal telephone service is made available. Internal telephone service also may be arranged to permit dictation over a telephone to recorders located in a stenographic pool.

Telegraph service requires neither internal personnel nor facilities, unless volume warrants installation of telegraph or teletype equipment.

Incoming mail and parcel post must be received, opened, dated, classified, and distributed. Outgoing mail requires collection, stamping, and dispatch. As an organization grows, there is merit in centralizing these functions at a mail desk or mail room.

Messenger service for internal communication is desirable whenever the organization exceeds the size to permit easy personal contact of all staff. This service can be integrated with the mail function. It can also circulate magazines and books from the library, materials for duplicating, and materials to and from the files, thus saving time for the professional staff.

Library

In sizable consulting engineering firms, a centralized library and reference service is useful. It may be combined with current periodicals, files for catalogs, data, reports, and other engineering literature. To make a library functional, procedures are needed for indexing and for checking materials in and out. The library operation can be combined with other assignments. A full-time librarian is needed only in a large organization or one requiring considerable research.

Transportation

Consulting engineers travel to reach clients and projects. This fact necessitates advance reservations for rail, bus, and air transportation, hotel rooms, and rental automobiles. It may also involve assignment of company-owned automobiles, vehicles, or planes. These traffic responsibilities may be assigned to a clerk or secretary in order to relieve higher-paid personnel. This activity can be combined with other functions, for it is normally far short of a full-time operation.

Property

The property of a consulting engineer consists of furniture, office machines and equipment, engineering and transportation equipment. Although the investment in property is not great, it is sufficient to justify good administration.

Property management starts with the selection of equipment and includes its assignment and maintenance. To avoid loss, a system of property accountability is desirable. Adequate records, with description, date of purchase, cost, depreciation and disposition, are needed. As the number of items increases, it is helpful to identify them by attached property numbers.

Maintenance requirements for furniture are simple compared to other equipment. Service contracts with manufacturers may be used for servicing office machines. Sometimes engineering equipment is returned to the manufacturer for overhaul. Unless a large fleet of vehicles is operated, they will usually be serviced by a commercial garage. Similarly, airplanes will be serviced and maintained by a fixed base airport operator.

To manage property effectively, responsibility should be assigned to someone on a part- or full-time basis, thus assuring adequate attention.

Purchasing

Although purchasing is limited, there are continuing needs for supplies and equipment. Efficient purchasing can save money.

This responsibility should be delegated to one person— owner or employee—who serves as purchasing agent. Purchasing should be initiated by a requisition that, after appropriate approval, authorizes the purchasing agent to buy. Standing policies are desirable in guiding him on sources, competitive proposals, and quantities. His operations can be facilitated by a standard form for requisitions, purchase orders, and receiving reports. After purchase, the agent follows through, expedites delivery, and assures receipt. If he does not personally receive the material, the recipient should notify him of receipt and condition. Often purchasing agents then approve the vendor's invoice for payment. Like many other service operations in a consulting office, purchasing is not a full-time function and can be combined with other duties.

Personnel

Few consulting firms can justify a full-time personnel director; nevertheless, there are many phases of personnel administration that the owners should delegate to assure adequate attention. A part-time personnel director, owner, or employee, perhaps with a nonprofessional assistant, is a workable arrangement.

Personnel functions may include assistance in recruitment and selection of personnel, obtaining of performance ratings, indoctrination of new employees, administration of company insurance plans, scheduling of vacations, preparation of employee manuals, editing of a house organ, and supervision of company-sponsored recreational activities. These duties are in addition to the keeping of permanent personnel records.

Computers

Within recent years many consulting engineers have installed electronic computers in their offices; these are often operated as a service function, even though programming is considered an engineering task. The maintenance and operation of computers may be handled largely by technicians and service staff. If a computer is warranted, it will involve full-time personnel and hence can be set up as a separate function. Procedures are required to initiate programs, to establish priorities on the computer, and to set charges for the services.

Duplicating Services

Most engineering offices require much duplicating of drawings, specifications, reports, and other documents. If commercial duplicating services are available, a consultant may elect to send much of this work outside. However, many firms do the work themselves, either because commercial service is not available or for convenience or economy.

Duplicating services include reproduction of plans and drawings, printing of typed material for specifications and reports, and copying of letters or similar materials. Collating and assembling are also a part of the process. A firm need not be large to require the full-time services of at least one person for duplicating. If a firm provides its own service, a separate section is desirable.

Building Maintenance

Unless a consulting engineer rents office space with complete janitor service, he has some janitor or building maintenance problems. If the firm owns its own building, it has complete responsibility for maintenance and operation. Between these two extremes, partial service may be required. In certain situations, guard service may be necessary for proper security.

Organization

The personnel required to perform the various supporting functions will probably be in the range of $7\frac{1}{2}$ to $12\frac{1}{2}$ per cent of the total staff. This percentage will vary with the size and type of the practice and may be even higher if the firm requires such special functions as duplicating and computer operation. Therefore, even a moderate-sized practice will require several people for these functions. If four or five or more are needed, a separate department or section is justified. Such a subdivision has the following advantages:

1. It permits better supervision, control, and management of the important service functions.

2. It relieves the owners and key engineers of the administration of these functions.

3. It helps the development of an *esprit de corps* among the staff.

Since there are many ways in which the supporting functions may be organized, no generally accepted pattern exists. Each consulting engineer adopts the arrangement he believes best suited to his own situation. The selection of a competent office manager or head for the supporting service group is more important than organizational structure. Based on my own experience, it is preferable to place all supporting functions in an Office Group, except the operation of an electronic computer. Such an Office Group, managed by a head of the Group or office manager, can be divided into sections with the following functions:

Administrative: personnel, purchasing, property, and building maintenance.

Stenographic: stenographic and secretarial duties.

Communication: reception desk, telephone switchboard, files, mail, messenger service, and transportation function.

Duplicating: all types of duplication.

Accounting: financial and accounting matters.

The consulting engineer's requirements for plant and equipment are simple when compared to industrial and commercial concerns. Nevertheless, he needs adequate facilities and apparatus to achieve efficient operation and good morale.

chapter 16
Plant and Equipment

Basic Needs

The plant and equipment of a consulting engineer consist principally of an office with furniture, fixtures, office machines, and standard engineering paraphernalia. In addition, most consultants own some transportation equipment. Beyond these basic items, a practice may have specialized needs. Consulting engineers sometimes require laboratory apparatus, instruments and other testing equipment, electronic computers, duplicating equipment, or other items.

Policy Regarding Plant

Several principles can be used as guides to plant and equipment. Since payroll costs are the principal item of expense, physical facilities and equipment should be planned to save time and improve efficiency. The major element in a consulting engineer's practice is personnel. Therefore, physical facilities that help to develop pride and morale are a good investment. Professional recognition and stature in a community will be affected by the impression the public receives from the consultant's office and equipment. They should create a favorable impression—not elaborate or extravagant but adequate and respectable.

The Office

The center of activity of a consulting practice is its office, which is not only headquarters and official place of business, but the working place of most of the staff. Many factors enter into the selection of an office, size, arrangement, appointments, location, and ownership. In addition to these physical considerations, cost and economy must be controlled.

Perhaps the best way to select an office is to subject it to the same kind of engineering study and analysis that consultants apply to their clients' problems. Too often, this is not done and the consulting engineer operates from quarters he would never recommend to a client. This may be due to economic factors, particularly with new firms. However, it is often the result of procrastination and failure to appreciate the importance of adequate offices.

Size and Plan

The first consideration is the amount of space required to accommodate the present and anticipated staff. Some thought should be given to future needs, based upon the consultant's aspirations and plans toward expansion.

Total space requirements for personnel can be computed from unit-area requirements for different categories of employees and the estimated number in each group. Area requirements will differ for each of several groups, such as: owners and key personnel in private offices, other engineers in private offices, engineers at desks in large rooms, engineering aides, draftsmen, secretaries, stenographers, and clerks. Space requirements in each case depend upon type of furniture, number of visitors to be accommodated, and desire for compactness or spaciousness, factors often influenced by economics.

The number of private offices to be provided is a question that can be decided only by weighing such factors as prestige, privacy, and cost. In a recent remodeling of the author's firm, we designed fewer private offices but achieved a suitable degree of privacy by furniture arrangement. A special wall-divider unit provides both separation and storage.

To the space for personnel accommodations must be added areas for lobby, reception room, conference rooms, file rooms, library, storeroom, mailroom, and such items as laboratory, computer, and duplicating departments if these are to be accommodated. Finally, allowance must be made for hallways, aisles, and restrooms.

Conference rooms become doubly desirable, if use of private offices is minimized. A small conference room, adjacent to the entrance, is useful for brief visits with callers, thus keeping them out of working areas. Larger conference rooms are convenient for staff consultation. One or more well-appointed conference rooms are desirable for meetings with clients.

The physical arrangement of space deserves careful engineering study to obtain a workable layout. Such studies should give consideration to flow of traffic, both of staff and visitors, organizational structure, shifting and variable work load requirements in different portions of the organization, and similar related items. Flexibility and ease of expansion are important considerations for a growing organization.

Appointments

An office is more than floor space and arrangement; other items are important—decoration, lighting, ventilation or air conditioning, and accoustical treatment.

The decoration and furniture of an office create an atmosphere and appearance that affect both staff and visitors. Many consulting engineers are inclined to settle for drab and uninspiring quarters. If we wish to appear as competent and progressive professionals, we must reflect it in offices that are modern and attractive.

Good lighting is a must in any consulting engineer's office because all staff members use their eyes constantly for reading or close work. Modern fixtures designed for proper illumination intensities not only improve working conditions but enhance the appearance of the office.

Adequate ventilation or air conditioning is very important because of the high density of people in drafting rooms and other open working spaces. Good ventilation creates a more healthful climate and overcomes the annoyance resulting from smoking. Air conditioning can be justified on a strictly economic basis in any office where summer months bring high temperatures and high humidities. Only a small increase in working efficiency is required to pay the operating costs and fixed charges on adequate air conditioning.

Finally, good office design requires accoustical treatment, particularly in the areas of high personnel density and activity. It improves working conditions and promotes the concentration essential to creative activity.

Location

This question involves both the selection of the city for the consulting engineer's office and its site within the city. Formerly most consulting engineers were in the larger cities. In the last few decades, however, there has been a strong tendency to locate in smaller communities. These smaller cities often offer advantages in living conditions, costs of operation, and caliber of nonprofessional staff. The former advantage of the large city in respect to transportation and communication has been lessened by the telephone and the airplane. However, no formula can be given for deciding the community where a consultant locates. Quite obviously it depends on his personal desires, and upon the type of service he offers and the clientele he serves.

Once the city is chosen, the location within it depends upon several items, chief of which is the availability of suitable office space. A consulting engineer need not limit his considerations to the heart of the business district. Normally he has only a limited contact with the commercial district of the city in which he locates. His clients rarely visit his office and then usually by appointment only. The majority of callers at a consultant's office are representatives of contractors, manufacturers, or suppliers, who will seek out the consulting engineer wherever he is located. For all of these reasons, many consultants, even in large cities, select outlying or suburban locations for their offices, particularly if they construct their own office buildings.

Ownership

Although most consulting engineers start in rented offices, there is an increasing trend toward ownership of quarters. The 1957 survey of *Consulting Engineer* indicated that 25 per cent of consulting firms own their offices. There are several reasons for ownership. Sometimes acceptable rental space is not available. Sometimes it is a matter of economy. Many firms have built offices to get away from the business district or to provide for anticipated expansion.

However, most offices are probably constructed because the consulting engineer wants facilities that are especially suited to his operation and that provide better working conditions and create a more favorable impression on his clientele and the public. Whatever the reasons, the trend toward ownership of office space by consulting engineers is likely to continue.

Furniture

Most of the furniture requirements for a consultant's office are the standard items used in other business offices, desks, tables, chairs, bookcases, files, and so forth. Special needs are mostly confined to drafting furniture and drawing files.

In selecting office furniture, the first decision is on decor or style of furniture. Here, the consulting engineer has a wide choice, ranging from conservative, conventional design to modern and streamlined styling. Similarly, he has a choice of wood or steel furniture. Naturally his furniture will be chosen to harmonize with the style of interior decoration he prefers and to fit with the dignity of a professional office.

It is desirable to develop a fairly standardized list of furniture for individuals of various classifications, including (1) owners and principals, (2) key employees in private offices, (3) engineering supervisors, (4) design engineers, (5) engineering aides, (6) draftsmen, (7) secretaries, (8) stenographers, and (9) clerks and others who must be equipped. The type, size, style, and quality of items will probably vary for the different classifications.

Special furniture requirements include: drafting tables, plan racks, drawing files, and a large amount of storage space for books, catalogs, specifications, and similar items.

The choice of drafting tables is extensive, ranging all the way from a stand-type drafting board at which the draftsman is seated on a posture chair to the large table or desk-type arrangements providing storage space and work area beside the drafting board. In addition to a drafting board, draftsmen need tables or desks to store reference materials, instruments, and books.

Engineering offices handle a great deal of paper in the form of plans and drawings. Hence, not only draftsmen but most engineers can advantageously use table space in addition to desks for such reference materials. In addition, racks are required for storage of sets of plans. We have found definite operational and appearance advantages in a storage unit matching the divider bookcase units mentioned above. Large flat or vertical hanging files are also needed for storage of vellums and tracings.

Office Machines

Most of the office machines required in a consulting engineering office are similar to those used in other offices. The comments below touch upon a few aspects of selection of office machines for consulting offices.

Typewriters. Electric typewriters are advantageous since they produce neater type, more uniform shading, and give good performance when numerous copies are required. Electric typewriters can also be operated more rapidly than manual typewriters. Long-carriage typewriters are handy for large tabulations or tables. Elite type will produce more characters to the inch than pica type. The new executive typewriters produce easily readable copy, for they utilize proportional spacing.

Adding machines and calculators. These should be viewed as labor-saving devices and an adequate number provided.

Dictating equipment. Such equipment is particularly useful because engineers, when working on reports, specifications, and even correspondence, must frequently interrupt dictation to make computations or review other materials. Dictating equipment is also well suited to transmittal of reports from the field to the office because of the nonbreakable, mailable discs, belts, or tapes.

Vari-typer. This useful machine can replace hand lettering on tables, charts, and drawings. It is possible to change quickly and easily from one style or size and type to another by inserting a new type plate. More than 600 different sizes and styles of type are available, and technical engineering symbols and characters can be obtained for use on the vari-typer.

Intra-office communications. Except in small offices, the telephone switchboard is required so that several incoming or outgoing calls may be handled simultaneously. The number of calls and stations will determine the kind and size of switchboard needed. "Intercom" systems are sometimes used to relieve the telephone switchboard of intra-office communications.

Accounting machine. As a firm grows, bookkeeping, posting, or other types of accounting machines may be necessary, particularly if extensive cost accounting is maintained.

Microfilming equipment. As files become crowded, microfilming often provides an answer to the dual demands of reducing files while keeping material for reference. It can be applied to drawing as well as general files.

Electronic Computers

These devices have recently become popular and are finding their way into the offices of many consulting engineers. In some places, groups of consulting engineers combine in a joint ownership and operation of computers.

Electronic computers are here to stay and will find increasing use in consultants' offices. They are efficient for handling a vast

amount of repetitive work, permit solutions to otherwise inde-
terminable problems, and in many cases increase the accuracy
of computations. However, the use of electronic computers in
engineering offices necessitates extensive re-examination of design and computation procedures. Electronic computers may be
purchased or leased. In view of the rapid progress in this field,
most consulting engineers lease apparatus so they can more
easily take advantage of future developments. The selection of a
computer for a given office is a complex decision requiring consultation with experts in the field.

Engineering Equipment

Requirements for engineering equipment vary greatly with
the type of service offered. It is not practical to discuss such
needs except to touch upon a few of the more common items.

Most firms conduct some survey operations and, therefore,
require surveying equipment. In this category are transits,
theodolites, plane tables, levels, range poles, chains, level rods
and associated items. Firms that supervise construction require
surveying equipment and, in addition, need such items as slump
cones, sieves for testing aggregates, and apparatus for making
compaction tests. Firms that make exploratory subsoil borings
by hand need soil boring equipment.

For field testing of apparatus and equipment, there is great
variety in the apparatus required. Electrical instruments may
include volt meters, ammeters, watt-hour meters, and resistance
meters. Mechanical apparatus may include pressure gages, ther-
mometers, and vibration meters. Chemical apparatus may be
required for field analysis of flue gas and fuels. Requirements of
testing equipment are extremely varied.

Those engineering firms that maintain laboratories have corresponding specialized needs.

Duplicating Equipment

Nearly all firms use some equipment for copying or duplicating typed materials. In addition, some firms maintain complete
equipment for all types of duplicating on drawings, reports, and

specifications. Firms located in cities with good commercial duplicating service need not undertake this complete function.

For duplicating reports, specifications, and similar documents, the engineer has his choice of processes, including stencil, spirit, and offset equipment. If the volume is moderate, stencil or spirit equipment may be adequate. With a higher volume and a desire for better quality of printing, the offset process seems advantageous. Our firm has used all of the above types but now concentrates on offset equipment. It affords the opportunity for changing scale while duplicating, thus permitting drawings of reduced size. It is also excellent for multicolored presentations. For full effectiveness, photographic and dark room equipment are needed.

Increasingly, engineers find advantage in simple copying machines that reproduce one or more copies of a letter or other opaque material; apparatus is available under a variety of trade names and will soon become common equipment for every stenographer or secretary who has occasion to copy materials.

Many firms have their own "blueprinting" apparatus, although today black line or blue line prints predominate. Such apparatus, offered by many companies, comes in a variety of sizes and types from which the consulting engineer may select the ones most appropriate.

If the consulting engineer finds it necessary or desirable to operate a full-scale duplicating department, he will need, in addition to duplicating equipment, such accessories as sorters, punches, shears, and storage units.

Transportation Equipment

A consultant must have a means of travel to projects and clients. He may meet these needs, in whole or in part, by owning a fleet of automobiles or by leasing vehicles.

Where vehicle requirements are limited and there are no special needs for survey parties or field expeditions, the firm may elect to have employees use their own cars and pay them a mileage charge. However, employees may not want to use the family car for extensive travel and especially for the rough field travel involved on surveys. Thus, if a consulting engineer has extensive travel needs, he is likely to meet them, at least partially, by ownership or lease of vehicles. If the volume of travel is large, the operation of his own vehicles will probably be cheaper than payment of mileage. Thus, his gains may be both in economics and in good will.

The decision between ownership and leasing of vehicles from a company offering this service is an economic matter. By leasing a fleet the engineer releases funds otherwise tied up in automotive equipment. This may be a controlling element if the engineer is short of capital; otherwise, the decision can be based on comparative economics and convenience.

Transportation from city to city can be accomplished by conventional automobiles. Special problems arise with respect to survey parties and other field uses where roads are poor and considerable equipment must be carried. Jeeps, pickup trucks, station wagons, and carryalls are sometimes used. Recently, we have standardized on station wagons, with equipment-carrier racks on top, for survey parties and construction jobs except where terrain and road conditions are so severe as to require more powerful vehicles.

A few consulting engineers have their own aircraft. This decision must be made by comparing the advantages and economics of ownership against commercial or charter service. In dealing with the economics of private plane transportation, saving of time and convenience are big factors. However, the decision in each case will depend upon the circumstances and upon the preferences and prejudices of the owners toward air travel.

Other items of transportation equipment occasionally required are trailers for housing or field office use and boats for hydrographic surveys or explorations.

Property Management

Full value from plant and equipment are realized only if they are properly maintained and effectively assigned. The importance of property management has been stressed previously and management procedures have been suggested in Chapter 15.

Fiscal management, an important element in any enterprise, is often slighted by consulting engineers. It includes accounting and the provision and control of funds.

chapter 17

Accounting and Financing

Importance

Many beginners in the consulting engineering field overlook the importance of financial matters. They tend to concentrate upon interesting engineering problems to the neglect of fiscal affairs. It is easy to underestimate the amount of capital required to operate a consulting practice and to be satisfied with inadequate accounting procedures and costing methods. It seems desirable, therefore, not only to emphasize the importance of accounting and financing, but to suggest practices suited to the consulting engineer.

These suggestions should not be construed as an accounting system. The consulting engineer should obtain the assistance of a qualified accountant to develop a system that suits his needs.

Accounting

Goals

Accounting, which is the process of recording, classifying, and analyzing the financial transactions of a business, has at least four principal objectives:

1. To record financial transactions in such a way as to permit the preparation of periodic statements showing profit or loss and financial status. Profit or loss is usually shown by an operating statement and financial status by a balance sheet.

2. To record and summarize the information required to determine revenues and prepare billings to clients.

3. To analyze costs of operation both for specific engagements and for overhead.

4. To provide management with a tool, in the form of reports and analyses, to assist in control and management.

Elements of Cost

Before touching upon accounting procedure, it is desirable to examine the two kinds of costs involved in operating a consulting engineering practice, direct and indirect.

Direct costs are concerned with the execution of a particular job. To be so classified, a cost must be identifiable as applying to a specific job. Direct costs normally include:

1. Salaries for time on the engagement.

2. Identifiable expenses incurred specifically for the engagement (travel and living expenses, telephone and telegraph expenses, duplicating, postage, and supplies).

3. Payments for outside services used directly on the engagement.

Indirect costs, sometimes called "overhead" or "burden," include expenditures to permit the operation of the practice as a whole but not identifiable with a specific engagement. Indirect costs include:

1. Salaries for administration, sales, supervision, and supporting functions.

2. Payroll expenses, such as vacation, sick leave, health insurance, payroll taxes, and insurance and retirement programs.

3. Rent, utilities, supplies, repairs, postage, travel expenses, and such items.

4. Outside services, such as for auditors and attorneys.

5. Fixed charges on property used and money borrowed including depreciation, insurance, interest, and property taxes.

6. Income taxes.

As the line between direct and indirect expenses is not rigid, accountants have many options of classification. For instance, postage and stationery may be charged as a direct cost. However, the work involved in such an accounting may be too great to be warranted, so postage and stationery may be considered as indirect cost. Payroll expenses (item 2, above), being directly related to salaries, may be applied as a percentage to salaries and thus be considered a direct cost, or they may be treated entirely as an indirect cost. These and other options can be decided as a matter of preference.

Accounting System

Before an accounting system is set up, three basic decisions must be reached.

One of these concerns the use of a "cash" or an "accrual" basis. With a "cash" basis, income is recorded only as money is received, and expenses and salaries are recorded only as paid. With an accrual system, income is recorded when it is earned, even though it may not be invoiced or received for some time. Costs are recorded when incurred, even though they may not be paid for some time. Accountants generally recommend the accrual basis as giving the best indication of actual performance. It poses one difficult problem, however, the periodic estimation of fees earned but not billed. As a result, some consulting engineers adopt a modified accrual basis wherein costs are recorded as they are incurred, but income is recorded only as fees are invoiced.

Another decision relates to single or double entry bookkeeping. Accountants normally recommend the double entry system. However, for small one-man consulting operations, a single entry system on a cash basis may be adequate.

Finally, the consulting engineer must decide the degree of cost accounting he wishes. Does he want direct costs on each engagement or will he be satisfied with only total earnings and costs for a given period? For an organization of any size, cost accounting by jobs is not only desirable but mandatory to provide good cost control.

Classification of Accounts

No classification of accounts for consulting engineers enjoys wide acceptance. The divergent practices in use make difficult any relative comparison of performance by consulting engineers.

A classification of accounts is simply a listing of the ledger accounts that are maintained in an accounting system. Usually numbers are assigned to the individual accounts for identification and simplicity. An accounting system should deal with the following groups of accounts: (1) assets (current and fixed), (2) liabilities (current and long term), (3) capital, (4) income (operating and other), (5) direct costs, and (6) indirect costs (operating and other).

The suggested classification of accounts presented in Appendix C is similar to one I have found quite satisfactory. It is flexible and may be expanded or contracted to fit given needs. The data and figures to be recorded in the ledger accounts come from the sources discussed below.

Time and Payroll

A daily, weekly, or monthly time report is the source of time or salary charges. Such reports show the hours worked, usually classified by the appropriate job number or indirect salary account. From such reports, the accounting department prepares payroll and pay checks, having available the employees' pay rate and authorized payroll deductions. Using the time reports, it distributes payroll cost among the various jobs and indirect salary accounts. Accounting also maintains records for each employee of gross earnings, payroll deductions, and net payments.

Expenses

Whereas there is a single source of payroll data, there are several sources of original entry for expenses.

Petty cash. Usually a petty cash fund is maintained for small payments. The party responsible for it makes periodic reports of the amounts expended and the appropriate account to be charged.

Expense accounts. Those employees who travel usually submit an expense account listing authorized expenditures and indicating the account to be charged. The expense account, after proper approval and audit, serves as the basis for refund to the party incurring the expense and for distributing the cost to the appropriate jobs or indirect accounts.

Credit services. Credit cards and credit services are being used increasingly for such items as hotel, meals, air and rail transportation, rental automobiles, and gasoline. When credit cards are used, the billings therefrom serve as the source for payment and for distribution of the costs.

Purchase orders. Orders to purchase materials or equipment, accompanied by the supplier's invoices, are a source of original entry for expenses and capital investment in plant and equipment. They require audit, approval, and indication of the appropriate job number, indirect expense, or capital account.

Contract services. Frequently expenses result from a continuing contract with another organization. Contracts may cover rent, equipment maintenance, or similar items. Monthly billings against such contracts serve as a source of data. Utility services, such as telephone and electric, are of similar nature. All resulting invoices require audit, approval, and distribution to proper job or indirect expense account.

Other items. Besides the above items, there are other invoices covering expenses that arise from a commitment or request, membership dues, interest, insurance premiums, and the like. Invoices for such items, when approved and audited, serve as a basis for payment and distribution to the appropriate account.

Other Costs

Beside payroll costs and expenses discussed above, other indirect costs are:

Depreciation. Depreciation expense is usually computed once a year by applying predetermined rates to the original cost of property. Over-all rates applied to a group of items may be used or depreciation computations and records may be made for each major item. Depreciation rates may be average ones or those allowing more rapid write-off in earlier years, such as the "sum of years digits" or "declining balance."

Taxes. Property taxes are usually invoiced by the appropriate county or city office. Other taxes may include state and federal income, sales tax, unemployment compensation tax, social security taxes, and others. The amounts of such taxes normally result from a computation and report prepared by the engineer and submitted to the appropriate governmental agency.

Insurance. Many forms of insurance, that is, workman's compensation, public liability, and others, are based upon contracted rates applied to actual payroll. Hence, the actual costs are based upon analysis and audit.

Loss on accounts. When an account is found to be uncollectable, it may be written off as an expense. Alternatively, an annual amount, usually a percentage of gross revenue, may be expensed as a "loss on accounts" and credited to a reserve for "loss on accounts." Actual losses are then charged against this reserve.

Income

Operating income arises from the various engineering engagements. In addition, the consulting engineer may have "miscellaneous" income from other sources.

If the accrual method is used, operating income or revenue must be estimated at the end of each accounting period. If work has not been invoiced, estimates may be based upon the approximate percentage of work completed applied to the total expected fee. If, however, income is recorded only when invoices are prepared, then the determination of operating revenue is made directly from billings prepared for clients. Determination of revenue on an accrual basis is facilitated if related to job cost records as discussed below.

"Miscellaneous" income may include such items as blueprinting or duplicating done for outside concerns, forfeitures of deposits made by bidders on plans and specifications, gains on sales of capital equipment, and other items.

Subsidiary Ledgers

General ledger sheets are required for each of the ledger accounts included in the established accounting system, such as that suggested in Appendix C. In addition to the general ledger, detailed subsidiary ledgers are required in certain instances. In these, the general ledger becomes a "control" account, recording the total amounts shown on the supporting subsidiary sheets. Such subsidiary accounts are useful for the following:

> Accounts Receivable
> Accounts Payable
> Direct Job Costs (by jobs)
> Property Investments

For accounts receivable, a sheet is maintained for each client, to which is posted all billings and all collections, the net difference indicating the amount of the receivable.

For accounts payable, a sheet is maintained for each creditor, to which are posted invoices received and payments made. If a voucher check system is used, unpaid vouchers may serve as the accounts payable subsidiary.

The plant or property subsidiary requires a sheet for each individual piece of property or equipment showing its description, date of purchase, original cost and usually the annual and accrued depreciation.

Job Cost Control

Cost accounting requires a record sheet for each job upon which are recorded costs and earned fees. Billings to the client may also be indicated thereon, in which case the difference between the fees and invoices indicates work in progress.

The extent to which costs are broken down in such cost accounting is a matter of judgment. The simplest procedure is to use three categories, direct salaries, direct traveling expenses, and other identifiable direct expenses. However, if the organization is departmentalized, it is desirable to classify job costs by departments or functions, to provide means for budgeting and control.

Generally speaking, it is satisfactory to maintain job records using direct costs only. Such costs, compared to revenue, must provide an adequate margin for the overhead consisting of indirect salaries and expenses. Alternatively, certain or all indirect costs may be distributed, usually in proportion to direct salaries and thus included in job costs.

Statements

Periodic operating statements and balance sheets are needed, usually monthly but sometimes bi-monthly or quarterly. Occasionally, fiscal periods, consisting of four or five full weeks each, are substituted for months.

Operating statements normally show gross operating revenue for the period and for the year to date. From these are subtracted direct costs to obtain net revenue. The deduction of indirect costs gives net operating profit, which is adjusted for any "other" income or "other" expenses to obtain profit and loss before income taxes. The deduction of taxes (if a corporation) gives resulting net profit or loss. The balance sheet states the assets, liabilities, and capital accounts as of a specific date, grouped according to the adopted classification of accounts. Such statements can be prepared in accordance with usual accounting procedures. However, the consulting engineer will want to make certain the breakdown of direct and indirect costs is meaningful to him in controlling his operations. Operating sheets and balance sheets may be accompanied by statements of costs and revenues by jobs.

Variance Accounting

Cost distributions of salaries are normally accomplished by using an hourly rate obtained by dividing the gross monthly salary of each employee by the number of hours worked. This is accurate, but may result in a different hourly rate each pay period.

Cost accounting can be simplified by the use of standard hourly rates for various employees or for employees of various classifications. When such standard rates are used, the procedures followed are similar to those used in "standard cost" accounting, which is widely applied in industry. If the totals determined by standard hourly rates are different from the actual payroll, the difference is carried to a "variance" account, which is an adjustment to profit and loss. If this adjustment becomes appreciable, the standard rates may be changed to reduce the variance.

Mechanization

For the large consulting firm, the time required for manual distribution of costs and posting to ledgers becomes burdensome. Many types of bookkeeping machines, calculators, and other mechanized apparatus are available and should be used when the volume of work indicates they will be economical.

Fiscal Management

Capital Requirements

Money required for a consulting practice consists of working capital and fixed investment in plant and equipment. Gross working capital, or current assets, includes cash, accounts or notes receivable, work in progress, inventory of supplies, and prepaid expenses. Of these items, the major ones are accounts receivable and work in progress. Together they represent services performed in part or in whole for which the engineer has not yet received compensation. Their combined magnitude depends on many factors, such as terms and methods of payment for services, size and duration of average engagement, work load as it affects rate of progress, and promptness of accounting and billing.

My experience indicates that the composite total of accounts receivable and work in progress is seldom less than two months' average income. Unless tight control is exercised, it is apt to increase to three or four months' average income. Adding thereto an allowance for cash, inventory of supplies, and prepaid items, total working capital will probably range from 20 to 40 per cent of annual volume of work.

Fixed investments consist principally of furniture, fixtures, office equipment, engineering equipment, and transportation equipment. I have found that the original cost of fixed assets, excluding office building, will range between 7½ per cent and 12½ per cent of a firm's annual volume.

Combining working capital and fixed investment, total capital requirements might run from 27½ per cent to 52½ per cent of annual gross revenue. Using a probable range of 30 to 40 per cent for the total, it is evident that capital requirements are substantial, as seen in Table 13.

TABLE 13. Capital Requirements

Annual Gross Revenue	Amount
$50,000	$15,000– 20,000
100,000	30,000– 40,000
250,000	75,000–100,000
500,000	150,000–200,000
1,000,000	300,000–400,000
2,000,000	600,000–800,000

Source of Capital

The total capital requirement is partially offset by current liabilities consisting of amounts owing to employees and creditors for the immediate preceding period, accruals on taxes, and similar items. To maintain a healthy financial position, current liabilities should not exceed one third to one half of working capital needs. Such a range gives current assets/current liability ratios ranging between 3.0 to 1.0 and 2.0 to 1.0. The balance of required capital must come either from the owners as paid in capital or retained earnings or it must be borrowed.

Contributions by the owners is the conservative way to provide capital. However, most firms have situations where they wish to borrow money and hence must be concerned about credit standing. Borrowing is always difficult unless a substantial portion of the capital requirement has been contributed by the owners.

Loans

Generally speaking, there are two sources for loans: owners and outsiders. Sometimes one or more of the owners will be in a position to loan money to the practice when it is needed, rather than add to his investment. But often the owners will have invested all of their resources in the practice before they face the question of borrowing; this is likely to be true with a new firm. Hence, they must look elsewhere for a loan source.

Here they have two general areas, banks or individuals. Perhaps one of the owners has a friend or relative who, with confidence in the concern, is willing to loan money. If such a financial angel is not in sight, the other alternative is a bank loan. But banks are not prepared to risk or advance the capital that proprietors of small businesses, inadequately financed, are usually seeking. Bank loans, except on real estate, are of comparatively short duration and must be retired according to a predetermined schedule. Bank loans, therefore, are best suited to meet temporary cash requirements resulting from an abnormal amount of work in progress or some similar situation.

Loans may be secured or unsecured. The latter variety is available only to concerns and individuals whose credit standing is well established. Sometimes unsecured loans can be arranged if they are backed up by ample life insurance on the borrower. This may be arranged if the lender has confidence in the borrower, but desires protection against his death.

Loans may be secured by mortgages on property, by pledges of stocks, bonds, or other collateral. Sometimes a consulting engineer may borrow money on an assignment of the revenues expected from a particular contract. In other cases, he may be able to use accounts receivable as security.

To prepare for the day when loans may be required, it is advisable to establish banking arrangements early and to fully acquaint the banker with the nature of the practice and its present and future financial requirements. Moreover, it is helpful to establish a history of successful borrowing even though the amounts are small.

Collections

Reduction of the aggregate amount of accounts receivable and work in progress is the simplest means of reducing capital requirements. Careful attention is warranted to the following steps, which can be fruitful:

1. Negotiation of terms in engineering service contracts providing progress payments during the period of work.

2. Control of engineering work to reduce the total period of time required for a given engagement.

3. Acceleration of accounting and invoicing procedures to avoid delay in submission of statements to clients.

4. Attention to clients' procedures in approving and paying invoices to reduce delays.

5. Follow-up of overdue accounts, using repeat invoices, telephone calls, and personal contact.

The above steps hasten the conversion of work in progress into accounts receivable or accelerate the collection of accounts receivable. Both of these devices work to reduce capital requirements.

Occasionally, one has an account that cannot be collected by normal procedures. Most such situations arise from dissatisfaction on the part of the client or misunderstandings over the terms of the contract. Before extreme methods are taken to force collections in such cases, personal contact with the client is desirable to seek understanding and perhaps adjustment and compromise. Failing this, the consulting engineer may decide to turn the billing over to a collection agency or to an attorney for action.

Auditing

At the close of each fiscal year, financial statements will be prepared for the period. They serve as a record of performance and are also used to prepare tax returns for federal and state income taxes. In addition, they may also be used for determining bonuses, profit sharing, and distribution of earnings to owners.

An audit by an outside accounting or auditing firm is desirable. The cost of such audits, if made from year to year by the same firm, is not excessive. An audit gives assurance to the owners of the business that fiscal matters have been properly handled. It also is useful in establishing credit with a bank.

Budget Control

The use of budgetary controls has become widespread in industry and commercial enterprises, but few consulting engineers employ this device. It can be of great value to consulting engineers, even though their operations are variable and difficult to predict.

More consultants should acquaint themselves with budgetary procedures and their use in controlling costs. As a firm grows and owners cannot personally control all phases of the operation, budgeting becomes almost a necessity.

No doubt indirect costs are the logical starting point on budgetary control. It is not difficult to set quotas for the various indirect salary and expense accounts for an ensuing year or quarter. The persons responsible for controlling costs should participate in the establishment of these quotas. They should be related to the accounting system so that periodic operating statements can show performance against budget.

Budgeting of direct expenses is best accomplished on the basis of individual jobs. When an engagement is contracted and its scope defined, estimates can be made of direct salary and direct expenses by departments or functions. After review and approval, these estimates can serve as budgets or quotas and be used to measure performance. Here again, the individuals who will be responsible for control of costs should participate in preparation of budgets and should receive periodic reports comparing actual performance against budget.

Budgetary control is a useful tool as a part of fiscal management and cost control, even though most consulting engineers view it with alarm and distaste.

To stay in practice a consulting engineer must make a profit. Since fees are set by prevailing practice, costs of operation largely determine the profit.

chapter **18**

Costs and Profits

Profit Needed

It is scarcely necessary to emphasize the need for a consulting engineer to earn a profit if he is to continue in business. If he loses money consistently, his capital is soon depleted and he will have to close his office. Such is the inevitable result in a competitive free enterprise.

It is helpful, however, to emphasize the several uses for profits. Since most consulting engineering practices are started with marginal capital, profits are the prime source of the added funds required for expansion. Similarly, profits are the normal source for the cash reserves that every organization needs to tide it over lean periods. Finally, profits are the source of money to implement bonus, profit-sharing or other incentive programs for the employees of the consulting engineer.

Source of Profit

The profit of a consulting engineering operation in any period is the difference between gross income from earned fees and the costs of operation (direct and indirect). The only sources of increased profits are higher fees or lower costs or some combination thereof.

The consulting engineer may properly seek adequate fees; but, at the same time, he has the obligation to stress efficiency and economy so that fees remain within fair levels. No one can dispute the premise that fees should be adequate to permit profitable operation, but there is danger in too great emphasis on this subject. Desire for higher fees should never be allowed to obscure emphasis on efficiency and economy of operation.

In the long run, consulting engineers will be harmed if fees become so high as to price them out of the market. It must be remembered that a large segment of the consulting engineering profession is in competition not only with fellow engineers but also with clients' engineering staffs, organizations offering turnkey proposals, and with the engineering staffs of manufacturers, suppliers, and contractors. To stay competitive consulting engineers must give constant attention to their costs of operation.

Cost Standards

As a consulting engineer seeks to compare his direct and indirect costs with others to determine his relative performance, he is confronted with an almost total absence of reliable information. There are no authoritative sources at present to which he can turn for data useful in making such comparisons. In time, no doubt, professional organizations of consulting engineers will compile and publish such data. Until they are available, a consulting engineer must make his own analyses from his cost accounting records and such meager data as are available.

Direct Costs

There are no known guides or formulas for estimating direct costs for different kinds of engagements. Moreover, direct costs vary substantially because of local conditions and situations, even when jobs are seemingly the same.

On any given job, a consulting engineer can compare his direct costs, plus a proportional share of indirect costs against the fee obtained. This comparison will determine whether or not he has made an adequate profit. Such comparisons, although helpful, do have their weaknesses and are not an adequate guide as to the propriety of direct costs. Such a comparison deals with three variables: direct cost, indirect cost, and fee, any one or all of which may be out of line, even though a profit is indicated.

These uncertainties indicate the importance of cost accounting to analyze direct costs on each project and permit comparisons from job to job. Such cost accounting gives the consultant the best available tool for measuring and controlling direct costs. When such cost accounting is maintained over a period of years, a pattern will emerge that can serve for budgeting, control, and supervision of direct costs.

Indirect Costs

Some limited published data are available on indirect costs, and more will become available as professional societies study costs of operation. One source of such data is the recommended multipliers for determining fees by the time charge or the direct payroll plus a percentage method.

As stated in Chapter 5, multipliers, to be applied to direct salary, ranging from 2.0 to 3.0 or more, are recommended by various professional engineering societies. A report on a recent survey by the magazine *Consulting Engineer* comments as follows:

> While there is a terrific range of multipliers and percentages in use, several formulas are fairly popular. These are the simplified formulas in which the percentage figure is eliminated. The most popular are:
> Salary × 2
> Salary × 2.5
> Payroll cost × 2 (This approximates salary × 2.34.)
> There are some rather interesting extremes: one firm says it simply charges straight salary. This firm is either on its way out of business, or it is a one-man firm and the "salary" is high enough to cover all the costs of doing business. At the other extreme, one firm charges $4.50 for each $1.00 of salary. This firm should be

doing quite well if it gets much work on this basis. There are a substantial number of firms, however, that charge $4.00 and about an equal number at the other end of the line who charge $1.50 or less for each salary dollar.

Using the more popular of these multipliers (namely, 2.0, 2.34, and 2.50) and using profit percentages (before income taxes) of 10 per cent and 20 per cent, the computed indirect cost ratios are shown in Table 14 (as related to direct salaries):

TABLE 14. Comparative Indirect Costs

Multiplier	2.0		2.34		2.50	
Direct salary	$1.00		$1.00		$1.00	
Total fee	$2.00	$2.00	$2.34	$2.34	$2.50	$2.50
Profit	10%	20%	10%	20%	10%	20%
Profit	$0.20	$0.40	$0.23	$0.47	$0.25	$0.50
Direct and indirect cost	$1.80	$1.60	$2.11	$1.87	$2.25	$2.00
Indirect cost related to direct salary	80%	60%	111%	87%	125%	100%

The same survey of the magazine *Consulting Engineer* presents Table 15.

TABLE 15. How a Consulting Engineer Firm Disburses Its Income *

Item of Cost	Per Cent
Salaries, wages, and profits	74.2
Engineering materials and equipment	3.4
Office space	3.7
Office equipment and supplies	2.3
Advertising, publicity, and client promotion	2.0
Transportation	3.8
Dues and contributions	1.4
Insurance	1.4
Utilities	0.9
Taxes, other than income and payroll	1.1
Special services (legal, etc.)	2.0
Depreciation	1.4
Miscellaneous	2.4

* Based on total income before deduction of income or corporate taxes.

These figures are the average of all firms in the survey. There is little difference between large and small consulting firms, with the following exceptions:

Large firms spend a slightly higher than average percentage for office space and office equipment.

Small firms spend a higher percentage for engineering materials and special services.

Table 15 shows expenses other than salaries and profit to be 25.8 per cent of gross income. Most of these expenses are indirect costs. However, the bulk of "Transportation" (3.8 per cent), some portion of "Engineering materials and equipment" (3.4 per cent), and the portion of "Utilities" (0.9 per cent) that covers communication charges may be considered as direct cost. An arbitrary assumption is made, therefore, that 5.0 per cent of total income is direct cost and the remaining 20.8 per cent covers expenses included in indirect cost.

The approximate total indirect cost may be determined from

TABLE 16. Probable Indirect Costs Computed from Table 15

Item of Cost	Per Cent	Per Cent
Total fee	100.0	100.0
Less profit	10.0	20.0
Total cost	90.0	80.0
Less direct and indirect expense *	25.8	25.8
Total salaries	64.2	54.2
Less indirect salaries †	14.8	12.5
Direct salaries	49.4	41.7
Indirect cost—total		
Salaries	14.8	12.5
Expense ‡	20.8	20.8
Total	35.6	33.3
Indirect cost as per cent of direct salaries	72.1	79.8

* Total of expenses except salaries, wages, and profits shown on preceding table.

† At 30 per cent of direct salary.

‡ Total expense of 25.8 per cent, divided 5.0 per cent direct and 20.8 per cent indirect.

the above data. The analysis shown in Table 16 assumes profit percentages of 10 per cent and 20 per cent before taxes, and assumes a 30 per cent ratio of indirect salaries (including vacations, sick leave, administration, and supervision) to direct salaries.

A study by Robert Bonney "Cost of Operation Survey," pub-, lished in the January, 1959, Consulting Engineers Council Newsletter, is shown in Table 17. It is based upon a survey of thirteen Oregon engineering firms, eight offering services direct to the owner and five engaged in interprofessional practice.

TABLE 17. Costs of Operation

Item of Cost	Max.	Min.	Avg.*
		Costs Reported	
1. Productive salary (base)	$1.00	$1.00	$1.000
2. Nonproductive salary	0.79	0.11	0.336
3. Holidays, vacations	0.125	0.02	0.084
4. Subtotal			$1.420
5. Payroll taxes	0.14	0.016	0.050
6. Hospital, med., life	0.06	0.002	0.013
7. Misc. benefits	0.02	0.01	0.003
8. Subtotal			$1.486
9. Prof. services	0.07	0.003	0.011
10. Eng's. materials, equip.	0.19	0.005	0.052
11. Office, util. tel. and tel.	0.24	0.054	0.107
12. Office equip., supplies	0.12	0.007	0.044
13. Adv. and promotion	0.076	0.002	0.022
14. Transportation	0.312	0.012	0.076
15. Dues and contributions	0.02	0.002	0.010
16. Insurance	0.03	0.006	0.011
17. Taxes, licenses	0.01	0.002	0.004
18. Depreciation	0.115	0.002	0.036
19. Miscellaneous	0.08	0.003	0.030
20. Totals	$2.78	$1.51	$1.889

* Weighted average.

A similar report of indirect costs of a Los Angeles consulting firm, published in the November, 1957, Newsletter of CEC, is summarized in Table 18.

TABLE 18. Study of Indirect Costs

Item of Cost	Cost
Direct Costs—Base Wage	$1.00
Holidays and vacations (18 days yr.)	0.07
Hospital, medical, life insurance	0.015
FOAB and CUI	0.0525
Liability insurance	0.02
Workman's compensation	0.0058
Miscellaneous benefits	0.02
Total direct labor cost	$1.1833
General overhead and administration based on total direct labor	
Accounting (legal and professional)	$0.0275
Advertising (dues, entertainment)	0.0455
Auto	0.0447
Bad debts	0.008
Depreciation	0.0127
Office salary	0.0715
Administration	0.1785
Sales	0.116
Supervision	0.0625
Rent (including utilities)	0.082
Telephone	0.0197
Travel	0.0402
Taxes and license	0.0054
Insurance and interest	0.0116
Miscellaneous	0.0178
$1.744 \times 1.1833 = \$2.06$ Total cost	$0.7436

Salaries of principals (four partners) are included in the above costs.

Reasons for Variation

The data presented in Table 14 through 18 on indirect costs show tremendous variation. The variation is so great as to raise doubts that the same basis of reference has been used in computing percentages of indirect cost. Many reasons contribute to these variations, including:

1. Lack of uniformity in accounting systems.
2. Wide differences in handling compensation of owners.
3. Variation in the cost of "readiness to serve" factor.
4. Appreciable differences in efficiency of operation and management, with respect to both direct and indirect costs.
5. Variations of costs because of geographical locations.
6. Differences in treatment of income taxes.

Each of these six items is now examined.

Accounting Variation

In Chapter 17 it was pointed out that the borderline between direct and indirect costs is an indistinct one and that accountants can properly select alternate treatment on many items.

One of the basic factors causing the apparent variation in indirect costs is attributable to differences in methods of accounting. Whenever an item is classified as an indirect expense, it tends to increase the percentage of indirect, both because the absolute amount of indirect cost is raised and because the direct cost (used as the base of comparison) is lowered. Cost comparisons will remain confused until uniform classifications are used for accounting or appropriate adjustments are made to achieve a truly comparable basis.

A significant accounting variation arises from the treatment of bonuses and profit-sharing payments to employees. If such items are added to base salaries in computing direct salaries, the result indirect cost ratio will be lower than if they are considered a distribution of profits or an indirect expense.

Owner's Salary

The percentage of indirect cost can be substantially varied by the method of compensation for the owners. If, for instance, owners are paid salaries that are distributed as direct and indirect costs, the results are substantially different than if the owners take no salary. In individual proprietorships and partnerships, the owners often do not pay themselves salaries. On the other hand, with corporations, it is almost certain that the owners will pay themselves generous salaries to achieve tax advantages. Even where owner's salary is treated as a cost, there can be a vast range with respect to salary levels. All of these factors affect the percentage of indirect cost.

Readiness To Serve

Consulting engineers have long contended that fees should include an allowance for their "readiness to serve." A client calling upon an engineer expects prompt service. To be in a position to give such service, a consulting engineer often maintains staff through slack periods. Salaries during such periods become an indirect cost, for they cannot be charged to a specific engagement. Indirect cost will also reflect associated overhead costs for staff not working.

The actual cost of such "readiness to serve" will vary greatly with different firms and types of practice. As the size of the firm and the diversity of its work rises, the cost of readiness to serve will normally decrease. In the author's experience, with a sizable, diverse practice covering several fields of engineering, the cost of readiness to serve has been a negligible factor. On the other hand, for some small organizations practicing specialties, it can be a substantial factor in determining costs.

The cost of readiness to serve is often overemphasized in an effort to achieve an adequate fee. It sometimes serves as an alibi for failure to maintain a more consistent and efficient operation.

Efficiency

Variations in indirect expense as related to direct costs can be closely correlated with the efficiency of the organization.

The more efficient the execution of surveys, studies, designs and supervision of construction, the lower the direct salary cost. Concentration upon efficiency in this area of operation may result in higher percentages of indirect costs, unless similar economy is achieved in indirect costs. I suspect the tendency in most engineering organizations is to give greater attention to the direct than to the indirect costs. It is probably easier to achieve higher efficiency in the performance of engineering work than in the administration and overhead costs.

Overhead is an insidious malady tending to grow out of all proportion to needs, unless continuous bold action is taken to restrain it. The usual concentration on engineering work often results in a slighting of efforts to keep overhead costs under control.

Location

The location of a consulting engineering firm has some effect on indirect costs, particularly with respect to office space and nonprofessional and technician salaries. The variations in these items among geographical areas of the country and among small, medium, and large cities are greater than the usual variations in salaries of engineers. Moreover, property taxes, property insurance, payroll taxes and insurance, gasoline taxes, and other items differ in various states and localities. All of these factors affect indirect cost ratios.

Income Tax

With sole proprietorships and partnerships, income tax is paid directly by the owners and is not normally considered an expense of the organization. On the other hand, a corporation pays income tax on its profits, and such payments will certainly be considered an expense. These differences can result in appreciable variation in the resulting computation of indirect expense. Such wide variations can be avoided if comparisons are made using costs and profits before income taxes.

Observations on Indirect Cost

The following comments are personal observations based on 25 years of experience with costs in my own company, during which we have kept extensive cost accounting records.

It is preferable to measure indirect costs as a percentage of direct salaries, rather than as a percentage of direct payroll cost (including vacations, sick leave, payroll taxes, and such). Direct salaries are exact in amount and readily determined, whereas the related payroll costs vary appreciably depending on company policy and experience. Direct salaries should include any normal incentive compensation paid to employees other than owners and managers. We accomplish this by using standard hourly rates based upon employee's approximate total annual compensation.

In determining salary costs, salaries of owners who are active in the practice should be included. This should be done even with individual proprietorships. Such salary levels should be related to the over-all compensation schedules of an organization. They should be reasonable and in line with what the owners would pay other engineers to handle the management and engineering functions carried by the owners. They should not include a return to the owner for the money he has invested or the risk he undertakes.

It wise to classify the so- called "out-of-pocket" expenses related to the execution of specific jobs as direct costs. Such items vary greatly from job to job and distort overhead percentages if considered as indirect expense.

An efficient and busy firm following the patterns outlined above should achieve indirect costs in the ranges shown in Table 19, provided the owners are active participants and each carry a full load of engineering and administrative responsibility.

TABLE 19. Ranges of Indirect Costs

Item	Indirect Costs as per cent of Direct Salary
Payroll costs including vacation, sick leave and direct payroll taxes	9 to 18 *
Indirect salaries including supervision, office services, accounting, sales and general administration, and including payroll costs on indirect salaries	25 to 40
Indirect expense including all items except income tax	20 to 30
Total indirect cost	54 to 88

* May exceed this figure with a retirement or pension program.

A mature, experienced, diversified organization should be able to control indirect costs within the range of 60 to 80 per cent of direct salary. In recent years we have budgeted and achieved a normal ratio of between 65 and 70 per cent of direct salary.

Profit Level

What is an appropriate profit margin for a consulting engineering practice? Obviously this depends upon the method of accounting that is followed with respect to owners' salaries. If they are considered a part of expense, the profit level may be less than otherwise. If they are paid reasonable salaries that are treated as expense, the margin must cover income taxes and compensation to the owners for the use of their investment and for the risks they take. The risks assumed in an investment in a consulting practice are not great compared to those in many other business ventures. Moreover, investment is not large compared to annual volume.

An examination of profit margins in other types of businesses is interesting. A report prepared by Dun and Bradstreet, inc.* shows the ratios of net profits expressed as per cent of net sales, as given in Table 20.

TABLE 20. Range (by Lines) of Net Profit after Income Taxes Expressed as % of Gross Revenue

Line of Business	Median, %	Upper Quartile, %
Retail (12 different lines)	1.01–4.06	1.38–11.10
Wholesale (24 different lines)	0.35–2.68	0.72–6.20
Manufacturing (36 different lines)	0.77–9.45	1.29–13.88

As the profit ratios shown are "after income tax," profit ratios before income tax are roughly twice the figures given in Table 20. From this table it will be seen that few business enterprises in these fields earn a profit exceeding 7½ per cent to 10 per cent after income taxes or 15 to 20 per cent before such taxes.

Governmental agencies that negotiate engineering service con-

* Fourteen important ratios in seventy-two lines of business—retail, wholesale, and manufacturing for 1957, Dun and Bradstreet.

tracts seem to aim at a profit margin before income taxes (but after reasonable owners' salaries) of not over 10 per cent.

It would seem reasonable to expect earnings of 15 to 20 per cent before taxes, which on a corporate basis means roughly 7.5 to 10 per cent after taxes. This assumes, of course, that owners receive reasonable salaries for their services, which are considered as an expense. Many consulting engineers do better than this, but many fail to come up to this level. This range of profit seem proper to reward the owners for their investment and risk, and to provide reasonable capital for expansion of activities.

Fees

As discussed in Chapter 5, the recommended fee schedules of various professional organizations are the starting point for determining satisfactory fee schedules. However, as has been pointed out, they are not entirely satisfactory because they are not based upon adequate cost analysis and they represent some compromise of judgment. Under certain circumstances they can be grossly unjust to a client or to a consulting engineer.

Intelligent fee determination requires adequate cost accounting to produce a record of cost factors that the consulting engineer can use to check the propriety of suggested fee schedules. Often he needs a higher fee than the minimum recommendation, but occasionally a lower one will be adequate to give a suitable profit margin.

Fees set by judgment, as is the case in most published schedules, tend to reflect average efficiency and average overhead costs. Therefore, a vital element in establishing fees is to analyze the resulting profit for various multipliers on direct salaries and various ratios of indirect cost. Such an analysis is presented in Figure 11 for the common range of multipliers and indirect cost ratios. The relationship may be expressed as:

$$P = \frac{M - (1 + IC)}{M}$$

where P = % profit before income tax
M = multiplier applied to direct salary to determine fee
IC = ratio of indirect cost to direct salary

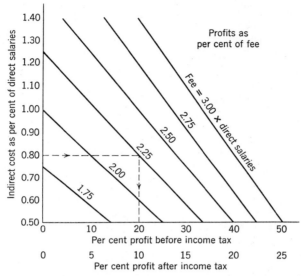

Figure 11. Chart showing common range of multipliers and indirect cost ratios.

The profit percentage on Figure 11 after income taxes assumes income tax of 50 per cent.

Supplementing the chart of Figure 11, Table 21 shows resulting profits for the more popular multipliers discussed earlier

TABLE 21. Profit Ratios

Multiplier	2.0	2.0	2.34	2.34	2.50	2.50
Direct salary	$1.00	$1.00	$1.00	$1.00	$1.00	$1.00
Total fee	$2.00	$2.00	$2.34	$2.34	$2.50	$2.50
Indirect cost, per cent *	70	80	70	80	70	80
Direct salary plus indirect cost	$1.70	$1.80	$1.70	$1.80	$1.70	$1.80
Profit, before income tax	$0.30	$0.20	$0.64	$0.54	$0.80	$0.70
Profit, after income tax †	$0.15	$0.10	$0.32	$0.27	$0.40	$0.35
Profit after tax, per cent	7.5	5.0	13.7	11.5	16.0	14.0

* Per cent of direct salary.

† Assumes income tax at 50 per cent.

(namely, 2.0, 2.34, and 2.50), using indirect costs of 70 and 80 per cent of direct salary.

It is my considered judgment that under normal situations a well-managed, efficient firm can show a satisfactory profit with 100 per cent markup on direct salaries. This means a multiplier of 2 times direct salaries as used in Table 21.

In addition to such fees for time expended, direct expenses for travel and living expenses, including duplicating, telephone, and telegraph costs, must be reimbursed or covered in the fee schedules. When fees are based on time charges or cost plus a percentage, such items are normally reimbursable. When fees are on a lump sum or per cent of construction cost basis, they must include an adequate allowance for such "out-of-pocket" direct expenses in addition to direct salaries, indirect costs and profit. Direct expenses vary tremendously from job to job, depending on distances traveled and other related factors. It has been my experience that on a given project they may run from 10 to 40 per cent of direct salaries excluding payments for outside professional services, such as soil borings, testing, and laboratory work.

Of course there are firms that require more than a 100 per cent multiplier on direct salary, or even on direct payroll cost, because of a high "readiness to serve" factor, an extensive amount of research, an expensive location, or other special circumstances. I believe, however, that any firm that is not showing a satisfactory profit on fees based on a 100 per cent markup on direct salaries, plus direct expenses, should carefully scrutinize its methods of operations and seek greater efficiency and reduced operating cost.

When it comes to establishment of per diem or hourly fees for owners or principals of a firm, it may not always be desirable to use the same basis as for employees. Often the owners' salary includes a return on investment and compensation for manageagement and administrative functions beyond his value to a client on a strictly engineering basis. Therefore, a judgment factor is in order in determining an appropriate per diem fee for owners.

Cost Reduction

Cost reduction is a continuing and important function that must receive attention lest costs become excessive and encroach upon legitimate profit margins.

As he seeks to give ever better services—a laudable goal—every consulting engineer is confronted with rising direct costs. Someone is always suggesting an additional study, another drawing, a specification revision, or a different analysis. Each of these may result in better engineering but they also increase direct cost. Similarly, the tendency to increase supporting and overhead functions creates a strong pressure in the direction of rising indirect costs. This is a condition typical of every business organization.

Every consulting engineer should be cost conscious and should strive for lower direct and indirect costs. Lest this emphasis be interpreted as a call for lowered quality of engineering, the full intent of cost reduction should be made clear. Cost reduction, if properly done, will not diminish the quality of the end product: engineering judgment, reports, designs, plans and specifications, and supervision. On the contrary, cost reduction should improve quality by stressing the important and minimizing the unimportant. Basically, cost reduction, accompanied by cost control, seeks to do each operation more efficiently and to avoid unnecessary or waste effort.

Continuing Need

Consulting engineers are confronted by several continuing needs, all of which are intimately related to the subject of this chapter—cost and profits. Four such needs are restated for emphasis:

1. More intelligent determination of fees to assure that they are equitable to both the consultant and to the client.
2. Adoption of a uniform classification of accounts.
3. Maintenance of adequate cost accounting records and controls.

4. Continuing efforts toward reduction of both direct and indirect costs.

These are objectives that consulting engineers must seek both individually and collectively.

*Success in consulting engineering
results from a judicious mixture of
good engineering, sound management, and enlightened professional
attitudes.*

chapter 19

Business or Profession

The Question

Part One of this book dealt with the role of the consulting engineer as a professional practitioner and his relationships to clients. Part Two has been concerned with the business and management aspects of a consulting practice. These two discussions may prompt the question: is consulting engineering a business or a profession? Actually, both elements are present. Certainly giving engineering assistance and advice to a client is a professional act. But it is equally true that any organization that seeks a profit as it handles hundreds of thousands or millions of dollars a year and employs tens or hundreds of people is a business. Hence, the challenge is not to debate the question but to mix sound management and professional attitudes, so as to set the stage for good engineering.

This chapter attempts to summarize the elements of both sound management and of professional attitudes, which have been discussed in previous chapters, and relate them again to the end objective of good engineering.

An authority on management, William H. Newman,* divides management, or administrative action, into five steps, namely, planning, organizing, assembling resources, directing and controlling. These five steps, unusually applicable to a consulting practice, are each discussed below.

Planning

Planning, as it relates to management, entails the determination of goals, the establishment of policies, and the setting up of procedures. These may apply to the operation as a whole, to a subdivision thereof, or to a special project.

The over-all goals of a consulting engineer concern location of office, geographical area to be served, types of services to be offered, and the size of practice and organization desired. Supplementing such general goals, financial plans are necessary for fees, annual revenue, costs, profits, and capital requirements. Policies and procedures relate to a host of subjects ranging all the way from personnel matters to drafting standards. Policies and procedures are of two types, those that are permanent or standing and those that are developed to meet special situations.

Planning is also required on every engineering engagement undertaken. Here it involves decisions on the work to be done, schedule to be met, manpower required, and the cost of performing the work. Such planning is the basis for the second step, "organizing."

Organizing

This step in the usual consulting engineering practice is concerned largely with organization of the over-all operation, including legal entity, organization structure, personnel management, methods and techniques, and service functions. All of these relate to the broad function of organizing a practice.

Beyond this, organizing is required on each engineering engagement that involves any departure from the normal pattern.

* William H. Newman, *Administrative Action*, Prentice-Hall, Englewood Cliffs, New Jersey, 1951, p. 4.

Such departures may involve a special task force or unusual arrangements for liaison and communication or other special conditions.

Assembling Resources

Resources required by the consulting engineer are personnel, money, plant, and equipment. Of these, personnel is by far the dominant item. However, adequate financial resources and plant and equipment must also be marshalled.

Since personnel is the all important resource to be assembled, it demands the highest priority. The process of recruiting is never-ending and time consuming, and selections will not always be satisfactory. One cannot emphasize too strongly the desirability of personal attention on the part of the owners to provide an adequate number of competent engineers and technicians.

Directing

Once the activity is planned, organized, and the necessary resources gathered, the next step is direction. The best-laid plans will fail unless management adequately directs it composite staff toward the planned goals. Direction is an essential step, for without it there is chaos. Good direction requires instructions that are reasonable, workable, and, above all, understandable. It also requires follow-up to assure execution.

A vital part of direction is coordination because without it the various parts of an organization may go separate ways. Coordination is particularly important in consulting engineering. It can be obtained partially from a proper organizational structure and partially from adequate communication, but it also requires close direction and supervision.

Control

The fifth step, control, is the process by which management sees that plans are executed or, if they cannot be carried out, are modified. Control is integrally tied up with other aspects of administration, particularly those of organizing, supervising, and directing.

Usual elements to be controlled are quality of engineering work, schedules of completion and costs, both direct and indirect. Quality control on engineering work is most difficult and is accomplished in three ways: first, by using competent and experienced staff; second, by assuring that all steps, such as checking, review, and approval, are performed; and third, by final scrutiny of the work by key personnel.

Control of schedules and costs is more direct. For control to be satisfactory, management must set standards at appropriate points to measure performance. Good control requires accurate and prompt reporting of performance for comparison against such standards. Control also involves mechanics for rectifying or correcting situations where performance does not match standards.

Emphasis

The need for sound management in a consulting engineering practice cannot be over-emphasized. This is true not because management is more important than professional attitude and engineering judgment, but because so many consulting engineers are prone to overlook or neglect good business practices. They are too apt to concentrate on problems of engineering and to neglect giving management problems their due attention, or they completely fail to understand management practices. It is my observation that more consulting engineers limit their success by poor management, including sales and human relations, than by poor engineering work.

Professional Aspects

Now, to look again at the other side of the coin, the professional aspects of consulting engineering. We must never forget that the consulting engineer performs a professional service and should be guided at all times by the highest ethical concepts. He should never allow business matters, profits, and costs to dim his professional attitude and conduct.

A professional attitude does not come about simply by discussion; rather, it is earned by the conduct of consulting engineers. It will be achieved only if decisions and actions are professional within the organization as well as with the public, clients, and others. The fundamentals of the professional approach are demonstrated in competition for engineering engagements and in relations with public, client, other engineers, and the engineering profession. They are summarized here for further emphasis.

The Starting Point

Professionalism begins at home. Unless the consulting engineer demonstrates a professional attitude within his own organization, he is unlikely to do so outside his office.

The importance of professional recognition for engineering employees has been fully emphasized. This is the best means of assuring a professional attitude in the dealings with clients and others. Moreover, all consulting engineers should strongly encourage and support engineering registration among their staff. It is the accepted method of distinguishing the professional engineer from the horde of other individuals calling themselves engineers.

Attitude toward Public

The consulting engineer will recognize his responsibility to the public and to mankind, placing it above personal gain, and using engineering knowledge and skill to benefit humanity.

While he strives to develop a profitable consulting practice, he will seek to be a good citizen and to assume his responsibility in the important affairs of city, state, and nation. The consulting engineer has much to contribute in these areas because of his understanding of the industrialized civilization and economy, which are largely the product of his mind and efforts.

Attitude toward Client

The consulting engineer, as a professional, will render faithful service to his client and will honestly represent the client's interests.

The consulting engineer will also encourage the client to think of him as a professional man, using their many contacts to demonstrate the meaning of professional responsibility. He will show utmost respect for his client's interest and keep his relations on a truly ethical plane.

Attitude toward Others

The consulting engineer must also act as a professional in his relations with all others, whether they are contractors, suppliers, governmental employees, or others. His relationships with all of them should be governed by the highest standards of integrity, fair-dealing, and courtesy.

Attitude toward Profession

The consulting engineer displays his professional attitude in his relationships to the engineering profession as a whole. He should encourage the development of the engineering profession and make his contribution to the improvement of both technical and professional engineering.

Certainly he will want to participate with the owners of other consulting practices in Consulting Engineers Council, the private practice sections of National Society of Professional Engineers or other organizations concerned with the professional problems of consulting engineers.

Engineering

Finally, in this chapter, which started with the question: "Is consulting engineering a profession or a business?" I would like to point out that, although consulting engineering is both a profession and a business, it is fundamentally engineering. No consulting engineer will succeed unless the engineering studies, designs, and judgments turned out by his organization are sound. He can no more succeed without superior engineering ability than can a doctor without good medical knowledge.

Every idea and suggestion contained in this book is aimed at helping the consulting engineer do a better job of engineering. The suggested professional and business aspects of a consulting practice are a means to an end—better engineering for the client. They should never become ends in themselves.

Consulting engineering offers a challenging career to those who possess the qualifications for success and the desire for service in the most professional of all engineering fields.

chapter **20**

Career

Appeal

No field of engineering presents an appeal equal to that of consulting engineering. To this I offer my personal testimony.

I entered the consulting field by purchasing a half interest in a small practice in 1932, in the depths of the depression. This venture, fulfilling an earlier aspiration, came after six years of varied engineering experience, including a year of graduate study.

Since that first year, when the practice grossed only $6000, consulting engineering has given me a continuing panorama of new problems, new interests, and new challenges. This recurring newness, or variety, is one of the greatest appeals of the profession. Each engagement has some facet that is different, the client, the problem, or the location. Repetitive situations are rare.

As a consulting engineer I have had the highest type of professional opportunity. This has been true not only in engineering matters but also in the broader fields of economics and public welfare. As an advisor to many clients, I have found the professional relationship full and rewarding.

I have had the stirring satisfaction that comes from converting an idea into a completed and operating project, the thrill of a job well done.

I have made a host of friends among the clients, contractors, manufacturers, governmental agencies, and others with whom I have worked. Such friendships are no doubt the most satisfying reward that can come from any vocational activity.

I have had the joy of helping to build a sizable consulting organization. The combination of engineering, business, and professional problems has been invigorating. Financial remuneration, after a few rough years at the beginning, has been more than adequate.

During my years as a consulting engineer, I have never regretted my decision to enter the field. The long hours of hard work and the occasional disappointments and frustrations have never dimmed the deep satisfactions of the challenging and interesting role of a consulting engineer. If I had the decision to make over again, I would make it the same way.

Career

Expanding opportunities in the consulting engineering field will offer employment to ever-increasing numbers of engineers, technicians, and service personnel. As projects handled by consulting engineers grow in size and complexity, the staffs of these concerns must include persons of increasing competence and experience. Many professional engineers and skilled technicians will find challenging opportunities for careers in the consulting field. Similarly, there will be growing opportunities for engineers to become owners or part owners of consulting engineering practices. Those who respond to this challenge will do so either by advancement within existing organizations or by the establishment of new ones.

Requirements for Success

No doubt it is presumptuous to list prerequisites for success in any field because there are always so many exceptions. Shortcomings can be overcome by work, drive, interest, motivation, and luck. Nevertheless, there are certain characteristics that enhance one's chances of success in a consulting engineering practice. Therefore, I hazard the listing below of six items I consider necessary to success in this field:

1. Superior engineering talent.
2. Scrupulous integrity.
3. Skill in human relations.
4. Ability in administration.
5. Compatible home atmosphere.
6. Spirit of an entrepreneur.

Obviously, the sixth requirement—the spirit of an entrepreneur —is required only of those who aspire to ownership of a consulting practice.

A consulting engineer must be a good engineer. Even though he may employ outstanding experts in different areas, he must earn their respect in order to lead the organization effectively. Moreover, he must often demonstrate his engineering ability to clients and others. Thus, a good engineering mind, a good engineering education, and good experience are prerequisites.

Honesty and integrity should be taken for granted as requirements for success in any business field. They are a keystone in engineering, where poor design in a structure or a machine will show for all to see that the laws of nature cannot be warped by intrigue, slyness, or deceit. Beyond this, however, the consulting engineer is in a unique position of professional trust as he deals with his clients, contractors, and others. Therefore, scrupulous integrity is an absolute requirement for success.

A touch for human relations is also necessary, although, unfortunately, lacking in many engineers. Interest in physical sciences often does not coincide with aptitudes in the field of human relations. However, the consulting engineer deals with people almost as much as with material and physical problems. His clients are people, the contractors with whom he works are people, and his employees are people. Success in sales and management is quite dependent upon skillful handling of human relations.

Another important talent is the ability to administer and to delegate. Some engineers never achieve this and insist on doing most of the detail work themselves. When such an engineer enters consulting work, he should confine himself to a one-man type of organization; for if he seeks to develop a sizable organization, he must delegate authority and responsibility in both engineering and management matters. Unless he possesses such a talent for administration, he soon creates a bottleneck, retarding the progress of his own organization.

A unique need is a compatible home atmosphere. One drawback to consulting engineering is that owners and key personnel must do considerable traveling. Moreover, the professional demands on consultants are heavy, as with other professional people, and, despite careful planning, there are frequent overtime hours to meet clients' needs. To superimpose management responsibility upon engineering places a heavy load on key people. Unless there is understanding and cooperation at home, a consulting engineer may be confronted with an impossible personal situation.

The consulting engineer who becomes an owner is in business as his own boss and manager. He needs the spirit of the entrepreneur as well as self-confidence and drive. Without it, he will not enjoy the risks he must take and the decisions he must make, nor will he be happy in the trying process of establishing his firm and developing his reputation. Not only will he be unhappy if he does not possess this spirit, but he will probably be unsuccessful.

Financial Needs

Fortunately wealth is not a prerequisite for entry into the consulting field. Many start in a limited fashion, expanding their practices by plowing back profits. However, financial resources must be adequate to carry the operation until revenues and earnings are available. Thereafter, they should be sufficient to avoid financial crises and to permit the desired expansion.

Most new consulting engineers probably underestimate financial needs and encounter resulting stress and strain. Therefore, care should be taken to appraise needs correctly, in order to avoid later headaches. To be truthful, however, I suspect that many of us would never have started if we had correctly appraised the capital requirements. Equally true, though, is the fact that many new businesses fail because resources are insufficient.

Financial Rewards

Just as financial requirements in the consulting field are moderate, so, too, are financial rewards. This is not to say that the consulting engineer may not enjoy a very adequate income. Successful consultants are perhaps better compensated than any other group of engineers whose activities are confined to engineering. I am sure an income comparison will show them ranking favorably with employed engineers whose compensation is not partially derived from broad management responsibilities. Nonetheless, consulting engineering does not offer such opportunities for wealth as are found in most business, industrial, and financial fields. Opportunities for great profits are related to large turnover and volume. The gross amount of fees of even a sizable firm is small compared to other businesses. Moreover, there is a distinct physical limitation of the effective size of a consulting organization.

Persons seeking opportunities for substantial wealth will not find them in consulting engineering; but those who aspire to the challenging career of a consulting engineer need not hesitate for fear of inadequate financial reward.

How to Start

Most engineers who become involved in consulting work start as employees with an existing firm. Once, however, a decision is made to have a firm of his own, the engineer has three paths to a consulting practice. He may start a firm alone or with partners. He may purchase an established practice or a major interest in one, or he may advance as an employee in an established organization until he achieves a share in ownership.

Of these paths, starting a new firm is obviously the most difficult. I would never advise a young engineer to do so until he had good experience and reasonable financial reserves. Purchase of an interest in a going concern may offer considerable advantage, for a newcomer can gain experience without full responsibility. If a partnership is considered, all parties should carefully evaluate their relative contributions to the organization and their ability to get along together. A partnership is a very intimate association that will not succeed without compatibility and mutual respect. Working up in an established organization is a satisfactory approach only if there is real opportunity for promotion and an ultimate share in ownership.

What Pattern

Once one is a consulting engineer, there are numerous paths of development from which to chose. These alternates are concerned with the fields in which the engineer will work, the kinds of service he will render, and the area in which he will operate. In connection with such decisions, he must resolve the question as to whether he will aim at a general practice or become a specialist in a more limited field. Also, he must make decisions regarding the size of an organization to which he aspires and whether he wants to be a sole owner or have others associated with him.

Opportunity and luck often answer some of these questions. However, as they arise, the consulting engineer will want to give careful consideration to his resources and, also, to his preferences, aptitudes and interests. A consultant who wants personally to handle difficult engineering problems should aim toward specialization within a smaller organization. Expansion requires the owners to become increasingly concerned with sales and management problems.

There is no formula to determine a desired pattern of development, for it depends upon the judgment and wishes of the consulting engineer himself.

Service

Above all else, success in consulting engineering, like that in any other true profession, requires a sincere dedication to service. This is the very essence of the professional attitude; service to others before self interest. If professional service is placed first, other matters tend to take care of themselves, including even financial returns; and the career of a consulting engineer becomes most challenging and rewarding.

Sample Contracts

formal contract for complete services

THIS AGREEMENT, made and entered into this *1st* day of *December,* 1959, by and between the City of State, hereinafter referred to as the "City," party of the first part, and X, Y, Z Engineers, Inc., an Iowa corporation, party of the second part, hereinafter referred to as the "Engineers,"

That Whereas, the City of State now owns and operates a sewerage system, and

Whereas, the Engineers have heretofore prepared and submitted to the City a report entitled "Preliminary Report, Sewage Treatment Plant Improvements, City of State—1959"; and

Whereas, the City desires to retain the Engineers to provide complete engineering services on the project;

Now, Therefore, It is Hereby Agreed by and between the parties hereto that the City does retain and employ the said Engineers to act for and represent it in all engineering matters involved in the project. Such contract of employment to be subject to the following terms, conditions and stipulations, to-wit:

1. *Scope of Project.* It is understood and agreed that the scope of the project shall include the improvements recommended in the "Preliminary Report, Sewage Treatment Plant Improvements, City of State, 1959" as prepared by X, Y, Z Engineers, Inc.

2. *Plans and Specifications.* The Engineers shall prepare such detailed plans and specifications as are reasonably necessary and desir-

able for the construction of the project. The specifications shall describe in detail the work to be done, materials to be used, and the construction methods to be followed.

The Engineers shall obtain approval of the plans and specifications from the State Department of Health.

Duplicate copies of plans and specifications shall be submitted to the City.

3. *Advertisement for Bids.* After the City has approved the plans and specifications, the Engineers shall assist in the preparation of notice to contractors and shall provide plans and specifications for prospective bidders.

4. *Award of Contract.* The Engineers shall have a representative present when bids and proposals are opened, shall prepare a tabulation of the bids for the City and shall advise the City in making the award. After award is made, the Engineers shall assist in the preparation of the necessary contract documents.

5. *General Supervision.* The Engineers shall exercise general supervision of construction work and shall review and approve all manufacturers' drawings. The Engineers shall process and certify to the City all contractor's payment estimates.

6. *Resident Inspection.* The Engineers shall furnish a competent resident engineer and/or inspectors to supervise the construction of the work. Said resident engineer and/or inspectors shall be assigned to the project during such periods as are mutually agreeable to the parties hereto. Such personnel and their salaries and expense allowances shall be subject to the approval of the City.

7. *Tests and Final Inspection.* After the construction is completed, the Engineers shall perform such tests as are necessary to make certain that all equipment and construction fully complies with the plans and specifications. The Engineers shall make a final inspection of the work and shall certify its completion to the City.

8. *Plant Operation.* The Engineers shall supervise initial operation of the sewage treatment plant and shall instruct the City's operating superintendent in the proper operation of the plant.

9. *Records and Reports.* The Engineers shall keep careful record of all construction and, upon completion of construction, shall provide the City with a complete set of plans showing the final construction.

10. *Property Surveys.* The Engineers shall not be required, under the terms of this contract, to make property surveys necessary for acquisition of right-of-way or property. The Engineers shall, however, make all topographic and construction surveys.

11. *Time of Completion.* The Engineers shall complete the plans

and specifications within 90 days after date of execution of this contract.

12. *Compensation.* The City shall compensate the Engineers for their services by the payment of the following fees:

A. For surveys, preliminary plans and estimates, final plans and specifications, and general supervision of construction a percentage of construction cost in accordance with the following schedule:

The first $15,000 of construction cost—ten per cent (10%)
Next $285,000 of construction cost —six per cent (6%)
Next $400,000 of construction cost —five and one-half
per cent (5½%)
All over $800,000 of construction cost—five per cent (5%)

B. For resident supervision and inspection, a fee equal to the salary paid the resident engineer and/or inspectors, plus seventy-five per cent (75%) of such salary paid, and plus any expenses incurred by the resident engineer and/or inspectors in connection with the job and paid by the Engineers.

Construction cost is defined as the total cost of the project exclusive of the cost of engineering, legal service, land and right-of-way and City's overhead.

The fee shall be due and payable in the following manner:

The amount of two thousand dollars ($2000.00) previously paid for the preliminary report will be considered as part payment of the fee stated above.

Upon completion of plans and specifications, their approval by the State Department of Health and presentation of plans and specifications to the City an amount equal to seventy per cent (70%) of the computed fee in accordance with Schedule A above based on estimated construction cost and less amounts previously paid.

During the period of construction and proportionally with the progress of construction an amount equal to twenty per cent (20%) of the fee in accordance with Schedule A above based on contract construction cost.

Upon completion of the project and final inspection, an amount equal to the total fee outlined above based on final construction cost, less amounts previously paid.

The fee for resident supervision and inspection as provided under Schedule B above will be billed and payable monthly.

13. *Services Not Included.* If, after the plans and specifications are completed and approved by the City and the State Department of Health, the Engineers are required to change plans and specifications because of changes made by the City, then the Engineers shall receive an additional fee for such changes which shall be based upon their standard per diem fees.

14. *Assistants.* It is understood and agreed that the employment of the Engineers by the City for the purposes aforesaid shall be exclusive, but the Engineers have the right to employ such assistants as they may deem proper in the performance of the work, said assistants to be employed subject to the approval of the City Officials, and the services of said assistants are to be paid for by the Engineers.

15. *Assignment.* THIS AGREEMENT, and each and every portion thereof, shall be binding upon the successors and assigns of the parties hereto, but the same shall not be assigned by the Engineers without written consent of the City.

16. THIS AGREEMENT, is executed in duplicate.

IN WITNESS WHEREOF, the parties hereto have hereunto subscribed their names the date first written above.

CITY OF STATE

By___John Smith_____
Mayor

Attest:

___J. P. Jones_____
Secretary

X, Y, Z ENGINEERS, INC.

By___Herman Brown_____
President

letter contract for study and report

X, Y, Z ENGINEERS, INC.

Municipal Building
Weatherbye, Illinois

June 20, 1959

City of Plainsfield
Plainsfield, Iowa
Gentlemen:

Engineering Services
Report on Electric Distribution System and Switchgear
Plainsfield, Iowa

Submitted herewith is our proposal for services in connection with an engineering study and report on the electric distribution system and switchgear.

The engineering study and report will include the following items:

1. General field examination of primary system.
2. Preparation of base map showing the location and size of primary conductors, distribution transformers and disconnect switches.
3. Electrical analysis of primary system based upon approximations of load distribution and estimated future load growth.
4. Recommendations for general development of system including observations regarding alternate possibilities for future system development including:

   a. System voltage.
   b. Number and general location of main feeders.

5. Establishment of priorities for necessary work to bring system up to ultimate design condition.
6. Estimates of probable investment required for initial improvements, together with maps showing initial improvements to primary distribution system.

241

7. Estimates of probable investment required for future improvements, together with maps showing ultimate primary distribution system layout.

City of Plainsfield

Plainsfield, Iowa

8. Comments on condition and practices of present distribution system and recommendations as to improving conditions of sag, tree trimming, maintenance, etc.
9. Comments on generally accepted practices of secondary distribution line sizes, circuit lengths, construction and maintenance.
10. Comments on switchgear and recommendations, with cost estimates, on necessary rehabilitation or replacement of existing gear. Recommendations to include ratings and general arrangement of major components of plan offered.

A written report will be prepared and presented in person to you. This report will summarize our findings and recommendations which will serve as the basis for determining appropriations necessary for required facilities.

The report will be presented to and discussed with you within approximately one hundred twenty (120) days after acceptance of this proposal.

It is understood that you will make available to us all plans, records and other pertinent information from your files which will be of assistance to us in our work and will also provide the services of a lineman to assist us during the time required to make the field examination of the system.

Our fee for the services outlined above will be Two Thousand Dollars ($2,000.00) which will be due and payable upon presentation of the report.

This letter may be made a contract upon your approval by affixing the date of acceptance and the appropriate signatures in the spaces indicated below.

Respectfully submitted,

X, Y, Z ENGINEERS, INC.

By   W. C. Green
_____
President

Accepted this   25th   day of   June   , 1959

CITY OF PLAINSFIELD
PLAINSFIELD, IOWA

Attest:

Jonathon Black
_____
Clerk

By   John   Henry
_____
Mayor

appendix B
Description of Grades

recommended grades, duties, for pre-professional

Classification— Grade and Number	Grade I Pre-Professional	Grade II Pre-Professional	Grade III Pre-Professional
Duties and Responsibilities Scope of Position	Perform routine tasks requiring knowledge of engineering fundamentals related to a particular field of work; work under close and immediate supervision.	Perform assignments requiring a basic working knowledge of engineering fundamentals related to a particular type of engineering work, usually work under immediate supervision of direction.	Perform assignments requiring a basic application of engineering fundamentals to engineering work; under direction but not immediate supervision and having limited responsibility and choice of action affecting design construction.
Examples of Work Performed	Compile data; compute quantities; extend estimates; trace or make simple drawings and sketches; make and record observations and measurements.	Make surveys; make and check quantity estimates or detailed drawings, working from designs by others; inspect minor fabrication, erection, assembly or construction for conformance to plans and specifications; make routine tests and inspections of equipment, materials and processes; set up or operate ap-	Select and recommend procedure in design and construction investigations, research, other engineering projects and write specifications or reports for minor projects following established engineering practices or general instructions; perform higher grades of drafting, prepare technical reports and

244

responsibilities and qualifications and professional positions

Grade IV Professional	Grade V Professional	Grade VI Professional	Grade VII Professional	Grade VIII Professional
Perform engineering assignments under general direction with the requirement for responsibility and choice of action in making decisions and interpretations affecting design or procedures.	Perform particularly important engineering work requiring special engineering qualifications or attainments, and offering wide latitude for independent action and decision.	Plan, direct, and supervise the work of a major engineering unit, or division engaged in design, construction, research, investigation or other technical operations and usually confined to a particular branch of Engineering.	Supervise and direct with final administrative authority a relatively large engineering or research organization comprising several divisions.	Supervise and direct with final administrative authority a large engineering or research organization comprising major divisions.
Engineering design, select and determine procedure in design, research, surveys, investigations and other engineering practices; write specifications and engineering reports following established engineering practices or general instructions; plan, conduct and	Plan, direct, and supervise the design or construction of major engineering projects; supervise the preparation of specifications and contracts; undertake comprehensive research and investigations; supervise testing work of importance; be responsible for ac-	To manage a small organization or a recognized major division of a larger organization engaged in design, construction, development, research or technical production and limited to a particular field of engineering; to assume professional	To determine and establish technical and administrative policies and procedures; to be finally responsible for all engineering research or technical operations of the organization.	To determine, establish and administer technical and administrative policies and procedures; to be finally responsible for all operations of the organization.

Continued on page 247

Description of

Classification— Grade and Number	Grade I Pre-Professional	Grade II Pre-Professional	Grade III Pre-Professional
Examples of Work Performed (*Continued*)		paratus or process equipment to obtain technical data; record technical observations and compile results as required.	recommendations on minor works.
Typical Position Titles	Instrumentman, Draftsman, Detailer, Junior Engineer.	Junior Engineer, Party Chief, Instrumentman, Checker, Quantity Es- timator, Maintenance or Construction In- spector; Engineering Draftsman, Laboratory Assistant, Technical Process or Equipment Tester; Assistant In- structor or teaching fellow in school of Engineering.	Junior Engineer, Senior Draftsman, Design Draftsman, Senior Inspector, Instructor in School of Engineering.
Minimum Qualifications Education			
Progressive Engineering Experience	None	One year	Two years
Engineering Status		Registered as Engineer-in-Training	
Ability and ca- pacity to plan, organize, super- vise, and coordi- nate technical work and to ob- tain cooperation from others			Some
General Adminis- trative Ability			

Grades

Grade IV Professional	Grade V Professional	Grade VI Professional	Grade VII Professional	Grade VIII Professional
report tests of materials, equipment and processes to obtain specified results; effectively recommend acceptance, approval or rejection of materials, fabrication or construction.	ceptance, approval or rejection of materials, fabrication or construction.	and executive responsibility for work of division; to give independent critical or expert engineering advice for executive action.		
Resident, Project, Office, Design, Test or Process Engineer, Chief Draftsman, Chief Inspector, Research Engineer, Assistant Professor in School of Engineering.	Project, Senior Office, Senior Resident, Senior Design, Senior Test or Process Engineer; Senior Research Engineer, Assistant Division Head, Associate Professor in School of Engineering.	Division or District Engineer, Production Engineer, Principal Engineer, Professor in School of Engineering.	Chief Engineer, Assistant Chief Engineer, Manager of Engineering, Director of Research Department Head in School of Engineering	Chief Engineer, Director of Research, Dean of School of Engineering.

Graduate with an accredited degree in engineering

Should have not less than four years of increasingly important engineering experience and for the higher grades added experience indicative of growth in engineering competency and achievement proportionate to responsibilities and duties involved.

Licensed Professional Engineer

Proven for ordinary engineering work.

Progressive ability, capacity and aptitude in either administration or in positions of research, teaching and scientific fields.

Some

appendix C
Suggested Classification of Accounts for Consulting Engineers

Current Assets:

111	Petty cash
112	Bank account
121	Accounts receivable (c)
121–01	Reserve for uncollectable accounts
122	Notes receivable
123	Accrued interest receivable
124	Advances to employees
131	Work in progress (j)
141	Prepaid expenses
142	Deferred charges

Fixed Assets

151	Investments
161	Furniture
161–01	Reserve for depreciation on furniture
162	Office machines and equipment
162–01	Reserve for depreciation office machines and equipment
163	Engineering equipment
163–01	Reserve for depreciation engineering equipment
164	Transportation equipment

164–01 Reserve for depreciation transportation equipment
165 Office building
165–01 Reserve for depreciation office building
171 Good will
172 Plans, drawings, and specifications

Current Liabilities

211 Accounts payable (c)
221 Notes payable
231 Accrued payroll
232 Accrued interest payable
233 Accrued taxes
234 Accrued vacation
235 Accrued sick leave
241 Employees deductions, withholding tax
242 Employees deductions, group insurance
243 Employees deductions, retirement program

Long Term Liabilities

261 Notes payable
262 Other

Capital Accounts (for Individual Proprietors or Partnership)

311 Capital, Owner A
312 Capital, Owner B
313 Capital, Owner C
321 Undivided profit or loss

Capital Accounts (for Corporation)

311 Common capital stock
312 Preferred capital stock
321 Earned surplus
322 Paid in surplus
331 Undivided profit or loss

Income

411 Income from jobs (j)
421 Other operating income
431 Miscellaneous income

Direct Costs

511	Salaries (j)
521	Travel expenses (j)
531	Duplicating expenses (j)
541	Telephone and telegraph (j)
551	Outside services (j)

Indirect Costs

611	Administrative salaries
612	Accounting salaries
613	Supervision salaries
614	Sales salaries
615	Service salaries
616	Vacations
617	Sick leave
621	Rent
622	Utility services
623	Office supplies and stationery
624	Engineering supplies
625	Postage
626	Repairs and maintenance
631	Travel expense—sales
632	Professional cards
633	Other sales expense
641	Library expense
642	Dues and licenses
651	Interviewing and recruiting
652	Contribution to employees group insurance
653	Contribution to employees retirement program
654	Other employee benefit program
661	Donations
662	Loss on uncollectable accounts
663	Administrative travel expense
664	Miscellaneous
671	Legal and accounting services
681	Depreciation
682	Insurance
683	Interest
684	Property taxes
685	Payroll taxes
686	Income taxes

Other Income and Expense

811	Income from duplicating department (including charges to jobs)
811–01	Duplicating department expense
821	Income from computer department (including charges to jobs)
821–01	Computer expense
831	Income from automotive equipment (including charges to other accounts)
831–01	Automotive expense

Notes

(c) Indicates a control account that shows total amount with detail carried on subsidiary ledgers.

(j) Indicates a control account that shows total amount with detail carried on job cost ledger sheets for each engagement.

Index